Dunn Public Library

Presented by

Alice Dantro

In Honor of

DAR
Daughters of the American Revolution
New Orleans Chapter
Spirit of '76

The Science Times

Book of

FISH

The Science Times Book of FISH

EDITED BY

NICHOLAS WADE

THE LYONS PRESS

Printed in the United States of America

Designed by Joel Friedlander Publishing Services, San Rafael, CA

10 9 8 7 6 5 4 3 2 1

Library of Congress Cataloging-in-Publication Data
The Science times book of fish / edited by Nicholas Wade.
       p.   cm.
   ISBN 1-55821-604-9 (cl)
   1. Fishes. I. Wade, Nicholas. II. Science times.
QL614.5.S35   1997
597—dc21   97-19471
                                                   CIP

Contents

Introduction 1

1   The Evolution of Fish

Biologists Debate Man's Fishy Ancestors 5
    MALCOLM W. BROWNE

Lake Victoria's Lightning-Fast Origin of Species 10
    CAROL KAESUK YOON

Reading the Record in Fishes' Ears 15
    JOHN NOBLE WILFORD

"Handedness" Seen in Nature, Long Before Hands 18
    MALCOLM W. BROWNE

2   Freshwater Fish

Guppies Hint How to Attract a Mate 25
    NATALIE ANGIER

They're Smart, for Fish, and a Model of Diversity 28
    NATALIE ANGIER

In the Amazon's Depths, a Strange New World of Unknown Fish 34
    CAROL KAESUK YOON

In the Quiet World of Fruit-Eating Fish, a Biologist Feels Too Alone 39
    MARLISE SIMONS

For the Male Molly, It's Not So Sterile an Affair 43
    NATALIE ANGIER

In Fish, Social Status Goes Right to the Brain 45
    NATALIE ANGIER

Biologists Embrace the Zebra Fish 49
    NATALIE ANGIER

Catfish Slime Has Healing Agents 56
    SANDRA BLAKESLEE

3   Freshwater Fish in Peril

Bizarre Parasite Invades Trout Streams, Devastating Young Rainbows 61
    WILLIAM K. STEVENS

PollutionThreatens World's Caviar Source 68
    Michael Specter

River Life Through U.S. Broadly Degraded 72
    William K. Stevens

Storm Swirls Over Aboriginal Salmon in Maine's Rivers .78
    William K. Stevens

Dwindling Salmon Spur West to Save Rivers 83
    William K. Stevens

Anglers' Gain Is Loss for Lakes and Streams 89
    Jon R. Luoma

Hatches and Wild Fish: A Clash of Cultures 95
    William K. Stevens

4   Oceanic Fish
The World's Deep, Cold Sea Floors Harbor a Riotous Diversity of Life 103
    William J. Broad

A Fantastic World Is Found in the Sea's Middle Depths .110
    William J. Broad

A Fish that Can Alter Its Body Size at Will 118
    Carol Kaesuk Yoon

The Odds of a Shark Attack 122
    William K. Stevens

In Dark Seas, Biologists Sight a Profusion of Life 125
    Carol Kaesuk Yoon

The Mystery of Why 25 Fish Species Are Warm-Blooded .131
    Catherine Dold

A Burrowing Fish Shapes the Sea Floor 135
    Walter Sullivan

5   The Sea's Threatened Fish
Creatures of the Deep Are Hauled to the Table 141
    William J. Broad

Debate Erupts over Peril Facing Ocean Species 146
    William K. Stevens

Throwing Back Undersize Fish Is Said to Encourage Smaller Fry 152
    Lindsey Gruson

Long-Line Fishing Threatens Fish and Albatrosses 157
    William K. Stevens

A Fish that Hails from the Age of Dinosaurs Faces Extinction 163
    Malcolm W. Browne

Despite Gaps, Data Leave Little Doubt that Fish Are in Peril 166
   DAVID E. PITT

Biologists Fear Sustainable Yield Is Unsustainable Idea 170
   WILLIAM K. STEVENS

The Terror of the Deep Faces a Harsher Predator 174
   WILLIAM K. STEVENS

Appetite for Sushi Threatens the Giant Tuna 180
   WILLIAM K. STEVENS

A Spate of Red Tides Is Menacing Coastal Seas 187
   WILLIAM J. BROAD

6   Shellfish and Others

Deadly Snails Take Pinpoint Aim with Diverse Toxins 195
   GINA KOLATA

Australian Navy Helps Endangered Giant Clam to Relocate 201
   NINA BICK

Roots of the King Crab 204
   NATALIE ANGIER

Squids Emerge as Smart, Elusive Hunters of Midsea 206
   WILLIAM J. BROAD

Scientists Close In on Hidden Lair of the Ocean's Fabled Giant Squid 213
   WILLIAM J. BROAD

Violent World of Corals Is Facing New Dangers 219
   WILLIAM K. STEVENS

Freshwater Mussels Facing Mass Extinction 225
   JOHN H. CUSHMAN, JR.

Earliest Samples of Royal Purple Found 229
   JOHN NOBLE WILFORD

Introduction

THE MAINSTREAM OF LIFE on earth, a dispassionate observer could claim, lies not on land but in water. The ocean is home to twice as many phyla, the major groups of living organisms, and may hold half of the planet's species. Terrestrial life is a late afterthought, a special niche for an assortment of weird species that like their oxygen almost neat.

It was a fish of some kind, perhaps an ancient forebear of today's coelacanths or lungfish, that set fin upon shore and began the watery vertebrates' parentage of the terrestrial kind.

But sentiments toward our fishy precursors smack more of cold disdain than the respectful interest that ancestors can usually count on. Fish inhabit a medium that requires scales and gills and other strange adaptations. Cold and clammy can't beat warm and furry in animal popularity contests. Nor do they make pets one can grow especially fond of, perhaps because of certain intellectual limitations in that fishy cranium.

Still, they are fellow vertebrates, and their family life bears a distinct resemblance to our own, as biologists are beginning to discover. Fish are generally harder to study than birds, say, because their homes are in a different medium, and this is especially true of fishes of the deep ocean.

The usual means of studying the ocean floor has been to drop down a scoop, or drag a sledge, and see what comes up—not a very comprehensive sampling method. A new generation of techniques, including robotic vehicles equipped with video cameras and other devices, is making the oceans much more transparent and accessible to biologists.

The new discoveries about salt- and freshwater fish occur at a time when fish populations in both habitats are under harsh and often unsustainable pressure.

Even though freshwaters throughout the United States are less polluted than they used to be, farming and logging have made many rivers inhospitable to fish. The native inhabitants of American rivers have been put under further pressure by the introduction of alien fish species. The aliens may take up residence as part of fish-stocking programs, or sometimes after liberation from an aquarium.

With ocean fisheries, the same sad story has been repeated with one species after another. A rich fishery is discovered; an industry grows up to exploit it but does so without restraint; the fishery collapses. Fishermen are left without a livelihood and the fish population is reduced to a size from which it may take years to recover. A current example of a fish now under heavy exploitation is the orange roughy, a deep-water species of New Zealand that grows slowly and can live to be 150 years old.

All these topics, from the evolution of fish to their biology and overexploitation, have been regularly covered in the Science Times section of *The New York Times,* which I edited from 1990 to 1996. Most of the articles here have been selected from the spate of fish news that occurred in that period. Though written for the purposes of a daily newspaper, they gain coherence from being brought together in a book.

Many of these articles appeared on the environment page of Science Times, edited by William A. Dicke. Mitchel Levitas of *The New York Times* oversaw the development of the book. The idea for the book came from Lilly Golden of Lyons & Burford. To the writers of Science Times belongs the credit for whatever diversion or instruction the reader may find.

—NICHOLAS WADE, Fall 1997

1

THE EVOLUTION OF FISH

The fossils of fish tell many tales, chief among them the story of how fish colonized the land and founded the lineage of terrestrial vertebrates that leads to humans.

But the lineage is still a little blurry near its beginning, leaving plenty of room for experts to debate just which branch of today's fishes should be regarded as humankind's closest cousins.

If it seems a large leap for a fish to morph into a human, the amazing cichlid fishes of east Africa's Lake Victoria offer a lesson in how fast fish can evolve when conditions are favorable.

"And in that place of all their wish/ There shall be no more land, say fish." Fish may hate the land for its support of fishermen, but the deed is done: The ancient spawning of air-breathing vertebrates cannot be revoked nor can the fossil record of evolution's strangest path be erased.

Biologists Debate
Man's Fishy Ancestors

MANY OF US FORGET that our ancestors were fishes, but among biologists the debate over exactly which class of ancient fish was closest to our forebears remains a hotly disputed topic. Was it the coelacanth or the lungfish that was more closely related to the family tree of four-limbed vertebrates that led to human beings? Some new laboratory and fossil evidence seems to favor the coelacanth. Far from settling the controversy, however, the latest discoveries have inflamed it.

In recent years, the preponderance of expert opinion has shifted toward the lungfish as the closest relative of the tetrapods, a group that includes amphibians, reptiles, birds and mammals, including ourselves. But new analyses of blood proteins and ear structures of present-day lungfish and coelacanths have convinced some scientists that the coelacanth is closer.

The first invasion of land by aquatic vertebrates nearly 400 million years ago was a momentous act in evolutionary history. It required the colonists to adopt new modes of locomotion, major changes in their organs of vision and hearing, and a new bellows apparatus for drawing oxygen from air, augmenting or replacing the gill system used by fish to draw oxygen from water.

No one can be certain which group or groups of fishes was the first to make the transition to land, or what their evolutionary pathways may have been. Nevertheless, the lungfish and the coelacanths each have their scientific partisans, while a few other scientists see more affinities between the tetrapod line and such ray-finned fishes as the tuna, carp and herring.

Lungfish are slim, eel-like fish with air-breathing lungs as well as gills, and four filamentary fins they use to feel out their surroundings. Fossil lungfish are found in sediments more than 400 million years old, and modern ones

5

live in freshwater streams and lakes in South America, Africa and Australia.

Coelacanths, with a similarly ancient pedigree, were believed to have died out about 80 million years ago, until a living coelacanth was caught off the South African coast in 1938. Its identification as a "living fossil" caused a scientific sensation. Since then, several hundred coelacanths have been caught in waters near the Comoro Islands, although none has survived the experience, and some scientists fear they are about to become truly extinct. These predators, about five feet long, have been photographed maneuvering into peculiar head-down stances with the help of their limb-like fins. They seek their prey on the ocean bottom and rarely, if ever, approach the surface.

In trying to decipher the evolution of tetrapods from fish, scientists face formidable problems. The transition from water to land occurred long ago, and various family trees suggested by the fossil record are so tangled that scientists acknowledge they may never be able to sort them out definitively.

The usual scientific approach is to examine every bony feature of all available fossils and to look for matches between different animals. If two animals share more "characters," or traits, than either animal does with the rest of the group under comparison, they are placed next to each other on a diagram called a cladogram, a tool of the classification system called cladistics, which groups animals and plants according to their similarities to each other. Animals with the fewest common characters are placed farthest apart on this type of diagram.

Cladograms merely show similarities between animal or plant groups and are not intended to portray evolutionary pathways, which are much more complicated. But evolutionary inferences are often drawn from cladograms, and these inferences are a major source of disagreement and contention.

When living animals are involved, comparisons become even more complex. They must take into account detailed features of organs, brains, nervous systems and such molecular components as DNA and proteins. Controversies have proliferated with each new type of trait that scientists add to their cladistic analyses.

Recent debate has focused on the work of two German scientists, Thomas Gorr and Dr. Traute Kleinschmidt, both of the Max Planck Institute for Biochemistry at Martinsried, Germany. They recently recalculated their earlier (and highly controversial) analyses of the blood hemoglobin of 522 marine and land-dwelling animals, and concluded that there is a close chemical match between one component of the coelacanth's blood hemoglobin and its corresponding component in tadpole blood.

The match, they reported, was not nearly as good between the coelacanth and an adult frog, however. The implication of this discovery, Dr. Gorr said in an interview, is that the metamorphosis of a tadpole into a frog not only recapitulates the gross physical changes the fish underwent as they evolved into air-breathing tetrapods, but also recapitulates the molecular changes from fish hemoglobin to tetrapod hemoglobin.

Dr. Gorr and Dr. Kleinschmidt concluded in an article in the February 1993 issue of *American Scientist* that "there is no reason to doubt that the coelacanth is the closest living relative of the tetrapods."

Disputing this view is another group of scientists, led by Dr. Axel Meyer, an evolutionary biologist at the State University of New York at Stony Brook. Dr. Meyer and the late Dr. Allan C. Wilson investigated the chemical ancestry of tetrapods from an entirely different direction three years ago. They based their comparisons on analyses of a special form of DNA found in mitochondria, the microscopic components of living cells that supply them with energy. This DNA, unlike ordinary DNA, is believed to be inherited only from the mother, rather than from both parents, and some scientists believe that mitochondrial DNA undergoes evolutionary changes at a steady, clock-like rate.

Dr. Meyer and Dr. Wilson showed that when the mitochondrial DNA of a frog was compared with the mitochondrial DNA of various fishes, including coelacanths and lungfishes, the best match was with the lungfish. "This result appears to rule out the possibility that the coelacanth lineage gave rise to land vertebrates," they concluded in a paper published by *The Journal of Molecular Evolution* in 1990.

In their rebuttal to this, Dr. Gorr and Dr. Kleinschmidt challenge the assertion that mitochondrial DNA is a better indicator of kinship than blood hemoglobin. Mitochondrial DNA evolves relatively rapidly, they say, and although it may be useful in establishing relationships between such recent species as modern elephants and extinct mammoths, it is unreliable for comparing very ancient animals with modern ones.

Dr. Meyer remains unconvinced. "The computer program they used in their analysis was never intended for the use to which they put it," he said in an interview. "Their methodology is wrong, and as far as I can tell, their reanalysis is no closer to the truth than their original work. If anything, it supports my lungfish hypothesis."

A different line of evidence supporting coelacanth ancestry has come from Dr. Bernd Fritzsch of Creighton University in Omaha. Dr. Fritzsch, whose research focuses on the evolution of hearing mechanisms, has studied the ears of coelacanths and lungfish, and compared them with those of land-dwelling tetrapods. He concluded that the detailed structure of a coelacanth's middle and inner ear is strikingly similar to that of the tetrapod, while the lungfish ear is like that of a shark.

The hearing mechanisms of all vertebrates must deal with a fundamental problem: the mismatch between the behavior of sound in water and its behavior in air, a phenomenon called impedance mismatch. In land-dwelling tetrapods, the inner ear is a water-filled cavity to which sound arrives from an air-filled middle ear. A structurally similar system, involving a bubble next to the inner ear, had already evolved in coelacanths before marine vertebrates began coming ashore, Dr. Fritzsch said.

Another type of new evidence that suggests coelacanth connections with tetrapods has

Dam Said to Threaten Ancient Lungfish

The Australian lungfish, *Neoceratodus forsteri,* which can use its one lung to breathe air, is one of the planet's strangest and hardiest survivors. It is one of the last remnants of the Dipnoi, a group of fish that has swum the world's rivers for more than 400 million years, managing to hang on as countless contemporaries, including the dinosaurs, died out and other creatures, like *Homo sapiens,* took their first steps. But conservationists say this species, now restricted to a handful of rivers in northeastern Australia, is being threatened in one of its last strongholds.

While a controversy continues to brew over the fish, the species' scientific value is in little doubt. *Neoceratodus* has fascinated researchers ever since it was discovered in 1870.

First thought to be salamanders, these creatures able to breathe air were later recognized as fish. Some scientists say they are the aquatic descendants of the lineage of revolutionary fishes that first flopped onto land and gave rise to reptiles, birds, amphibians—all the land-living vertebrates, including humans.

"It's a real case of a living fossil," Dr. William Bemis, an evolutionary anatomist at the University of Massachusetts at Amherst, said of the species, which is indistinguishable from its rock-hardened predecessors. Its only living relatives are the South American and African lungfish, both of which are considered much less primitive.

—CAROL KAESUK YOON, May 1996

come from an investigation of the oculomotor system, the brain and neural connections that control movement of the eyes. Dr. Chris von Bartheld of the University of Washington at Seattle reported recently that his study showed the oculomotor system of the coelacanth was closer than that of other fish to that of tetrapods.

Still, many if not most scientists remain unconvinced. Dr. Eugene S. Gaffney, curator of vertebrate paleontology at the American Museum of Natural History in New York, is among the doubters. "For many years I believed that the coelacanth was closer than the lungfish to the tetrapod line," he said, "but a few years ago I changed my mind. The cumulative evidence finally swayed me to the lungfish, and I haven't changed that view."

"We all must keep an open mind," Dr. Gorr said in reply to criticism by Dr. Meyer. "But why must some of our critics insist that only their own evidence is worth considering? I think scientists can be surprisingly bigoted."

—MALCOLM W. BROWNE, March 1993

Lake Victoria's Lightning-Fast Origin of Species

VIEWED AS AN EVOLUTIONARY TREASURE CHEST, crowded with hundreds of closely related fish species that can be found nowhere else in the world, Africa's Lake Victoria has long attracted biologists to its shores. But while scientists probed and prodded these cichlid fish (pronounced SICK-lid), in the vast watery expanse of this continent's largest lake, Victoria yet managed to conceal its most magnificent secret.

Now an international team of researchers, using remote sensing to probe the sediments at the lake's bottom, have found evidence that what is now Lake Victoria was a dry, grassy plain just 12,000 years ago. For evolutionary biologists, the implications are enormous. The discovery means that the 300 unique fish species that have been documented in the lake must have evolved in the unthinkably short interval since the current lake began to form, a geological instant ago.

"All these species, this whole universe of cichlid fishes, that all this could have evolved in 12,000 years," said Dr. Ernst Mayr, the Alexander Agassiz professor emeritus at Harvard University in Cambridge, Massachusetts. "As improbable as it seems—the facts force you to accept it."

Calling the conclusions, reported in the current issue of the journal *Science* "ironclad," Dr. Amy McCune, an evolutionary biologist at Cornell University in Ithaca, New York, said, "It's amazingly exciting. We're talking about rates of speciation that have not even been imagined." She added, "I think everybody is going to be surprised."

But now even more perplexing questions arise, as researchers ponder how so many species could have evolved in such a short period of time and what there is about cichlid fishes that makes them prone to such excesses of evolutionary speed.

"It's a world record, no question," said Dr. Axel Meyer, evolutionary biologist at the State University of New York at Stony Brook. In fact, these cichlids make evolutionary laggards of other groups much celebrated for their speed of evolution, like Darwin's finches on the Galapagos Islands. On those islands, Dr. Meyer said, fewer than 20 species of finches have evolved over four million years or more. In contrast, Lake Victoria has churned out more than 10 times as many cichlid species in less than one-hundredth the time.

The team of nine researchers that made the discovery had no particular scientific interest in the cichlids that swam alongside their boats. They were drawn to Lake Victoria in the hope of gathering clues about long-term climate changes from the lake's history of fluctuating water levels. The team was led by Dr. Thomas C. Johnson, director of the Large Lakes Observatory at the University of Minnesota in Duluth; Dr. Christopher Scholz, of the University of Miami; Dr. Michael Talbot, at the University of Bergen in Norway, and Dr. Kerry Kelts, at the University of Minnesota in Minneapolis.

Setting out in a fishing boat, the researchers used a method of remote sensing in which an air gun is discharged into the water, sending sound waves down into the ground. A series of sound detectors capture the returning echoes, which are used to determine just what lies below the surface, in a process known as seismic reflection profiling. Just as the sound waves of an ultrasound monitor can reveal the contours of a baby that lies below the surface of a pregnant woman's abdomen, the geologist's seismic profiling can reveal what sorts of sediments lay hidden beneath a lake bottom.

But instead of a long, continuous record of changes in the lake's size, which would help the climatologists, the researchers' seismic profiling indicated that the layers of lake mud seemed to disappear entirely not far below the surface. To see what they were detecting, researchers took cores from the lake bottom, sometimes extracting columns of sediment 30 feet long. Researchers were then able to decipher past events as they looked into the muddy accumulation of history.

Beneath the top layers of the sediment laid down by the current lake, they found a layer that held cattail pollen, which suggested a drier, more marsh-like setting. Just below that, they saw the really dramatic change their profiling had detected: a layer of grassland soil, complete with plant rootlets, grass pollen and dried, cracking mud. A remnant of a time of complete drying, the soil layer apparently stretches the entire reach of the lake, appearing even at its deepest point. Researchers used radiocarbon dating to date the refilling of the lake to around 12,000 years ago. Beneath the soil, the researchers say, they detected yet more lake sediments, deposited by even earlier incarnations of Victoria.

"The seismic data are very good," said Dr. Andy Cohen, a paleobiologist at the University of Arizona at Tucson. "This pretty well puts the nail in the coffin about whether the lake really dried up or not."

The new study corroborates data from the fish themselves as to how fast they evolved. In a previous study, Dr. Meyer and his colleagues used DNA from the fish to estimate how much time the Lake Victoria cichlids had to diverge into hundreds of different species. What they found was that the cichlids in Lake Victoria were extremely closely related, having diverged very recently; surprisingly few genetic differences were detected among them.

In fact, so few genetic differences had accumulated among the species that the lack of such data made it difficult to estimate how old this species flock, as these large groups of fish species are known, really was. Dr. Meyer said he and colleagues had been able to narrow down the range to between 200,000 and 14,000 years—a finding that is now nicely corroborated on its most rapid end by the newest geological work.

But the question arises: Could many cichlid species have survived the drying of Lake Victoria by hiding out in other lakes or rivers, recolonizing it when it formed again, which would make them much older than the lake itself?

The new study appears to rule out such a possibility. The topography of this very flat area is such that any rivers that feed the lake feed the deepest part first. Researchers say they can make a very convincing case that, with the deepest point dry, there was no satellite lake or pond in which a cichlid could have hidden.

As for the fish, they tend to be open-water fishes, not adapted to life in rivers or small ponds and unlikely to have survived well outside their lake. But researchers say it is theoretically possible that a handful of species, perhaps 10 at most, might have survived in very small, nearby bodies of water and might have recolonized the lake later. If so, that would have little effect on the count of species that did arise in Lake Victoria.

The last out for the skeptic is the suggestion that the species count itself is wrong, mistakenly elevated in the sometimes difficult business of delineating one species from another. But the researchers interviewed were in easy agreement that the more likely scenario for Lake Victoria, many of whose species remain poorly studied, is that the count of unique species that have evolved over the 12,000 years is higher than 300, rather than lower.

So if all these cichlids came to be in a blink of evolution's eye, how did they do it? "Well, if any group was going to do it," said Dr. Melanie L. J. Stiassny, curator of fishes at the American Museum of Natural History in New York, "it was going to be cichlids." Dr. Stiassny said there was abundant anecdotal evidence that these fish could evolve extremely rapidly. Fish caught in the wild and kept in captivity become markedly different in shape from the original stock in just 20 generations, she said.

The real puzzle, as posed by Dr. Mayr, arises because the most common way in which species are thought to evolve is by being isolated from one another geographically and slowly evolving to become two distinct species. How could groups of fish all swimming together in Lake Victoria ever be isolated enough to produce 300 species?

One hint, Dr. Meyer said, comes from the fact that cichlids are weak swimmers and have very specific preferences for certain kinds of habitats. By clinging, for example, to sandy bottoms and never crossing a rocky area to reach the next sandy stretch, he said, groups of neighboring cichlids might well have been able to be as isolated from one another as if they had been living in separate lakes.

But Dr. McCune said that even with such isolation, a mere 12,000 years was still very little time for so many species to be generated without something else being added to the soup.

What researchers have suggested is that cichlids may have a predisposition for rapid evolution in their mating preferences, which has led quickly to small populations isolated not by geography, but by a tendency not to interbreed. This is just the sort of change that could lead to the evolution of new species.

But while cichlids are the speediest of all, evidence is accumulating from both living and fossil species that fish, on the whole, seem to spin off new species extremely rapidly. Yet why fish might be so inclined is, at least for the moment, entirely mysterious.

"It's such a new question," Dr. Mayr said, "that nobody has really looked into it."

Meanwhile, in all of this excitement, Dr. Stiassny said, many of the cichlids are being driven to extinction just as biologists are figuring out what has been going on with these species flocks.

She said that Lake Victoria's edges were being deforested, that there was a good deal of runoff from agricultural lands bordering the lake and that much of the depths of Victoria were no longer even well oxygenated enough to maintain the species that once lived there.

Added to that problem is a kind of biological terminator, the Nile perch. A voracious predator introduced by humans into Lake Victoria to help the fishing industry, it has driven at least half of the lake's cichlid species extinct already and has laid many others so low there is little hope for their recovery. There are now only about 150 cichlid species left, and scientists studying the evolution of these fishes must often work on specimens saved in museums or research collections.

"It's so ironic," she said. "We discover this wonderful thing, this fabulous laboratory for the study of evolution, only to find that the whole system is collapsing."

—CAROL KAESUK YOON, August 1996

Reading the Record in Fishes' Ears

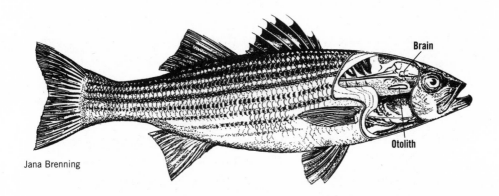

Brain

Otolith

Jana Brenning

Growths called otoliths vary with weather.

HEAVY-FOOTED ANGLERS know from hard experience that fish can hear their steps on the riverbank and dart away to safety.

The vibrations in the ground propagate through the water, impinge on bones of the fish and then travel to chambers inside the skull on either side of the brain and just behind the eyes. There the sound waves rattle tiny stones resting in beds of nerve hairs, which communicate signals of the disturbance to the brain.

This is the way fish hear, with two entirely internal ears each containing three pairs of calcium carbonate stones called otoliths, ear stones.

Now scientists are finding in fish ears a new and previously unsuspected function, not for the fish themselves but for research into the world's climate history. The otoliths from fish that lived long ago, they have discovered, preserve an amazingly detailed record of seasonal temperature variations over the past 200 million years. These stones can be, in effect, paleothermometers.

"What we've developed is an important new tool for studying global change," said Dr. Gerald J. Smith, a zoologist at the University of Michigan in Ann Arbor.

Of particular importance, analysis of otoliths promises to fill a serious gap in knowledge of climate history. The usual sources of information about past climates are fossils of one-celled organisms called foraminifera. Since these are found only in ocean sediments, they provide clues only to ocean temperatures. They may reflect general climatic conditions of a certain period, but only on very large time and geographic scales.

By examining otoliths from freshwater fish, Dr. Smith and his colleagues were able to learn about past temperatures in midcontinental regions, which are especially sensitive to seasonal climate changes. Thin growth layers inside the stones, like the rings of a tree, contain evidence of weekly, sometimes daily, changes in water temperature over the life of the fish, whether it died last week or a million years ago.

William P. Patterson, a graduate student in geology at Michigan, learned to read this record by studying museum collections of otoliths, some of which dated to the time of the dinosaurs more than 65 million years ago.

Stones from small fish are like grains of sand, barely visible to the unaided eye. Those from larger fish can be more than half an inch wide, about the size of small marbles. Each species produces otoliths that are identifiably different in size and shape.

Contrary to an opinion sometimes expressed, people do not have rocks in their heads, although their inner ears do contain some microscopic grains of calcium carbonate called otoconia. Bones in the inner ear, not the otoconia, serve much the same function for humans as the otoliths do for fish.

Many of the museum specimens of otoliths had been collected by geologists and paleontologists from layers of ancient sediments. Others were found in archeological excavations of American Indian settlements. The larger, more beautiful ones seem to have been prized "collectibles" among Indians in the Great Lakes area.

Mr. Patterson sliced open some of these stones to count the microscopic growth rings and analyze the ratio of certain oxygen components in the minerals of each ring. This ratio of oxygen isotopes is the key to using otoliths as paleothermometers.

"Because we can reconstruct average weekly temperatures with this technique, we can accurately track seasonality—the difference between winter low and summer high temperatures," Mr. Patterson said. "Seasonality is the most significant factor in determining which plants and animals can survive in a particular area."

Mr. Patterson, Dr. Smith and Dr. Kyger C. Lohmann, another geologist at Michigan, described their otolith research at recent meetings of the Geological Society of America in Cincinnati and the Society of Vertebrate Paleontology in Toronto.

Each day a fish synthesizes the calcium carbonate mineral known as aragonite, which includes oxygen from the surrounding water, and so gradually builds up the size of its ear stones. Most of the oxygen atoms in the mineral are in the form of oxygen-16 isotopes. A tiny fraction is oxygen-18, which is a slightly heavier atom because it contains two extra neutrons.

In a study of modern fish from the Great Lakes, particularly a large species of drum, Mr. Patterson and Dr. Smith found that the ratio of these oxygen isotopes deposited each day in the otolith corresponded to prevailing conditions in the lake.

As Mr. Patterson explained: "When the water and the fish are cold, the fish absorbs more of the heavier isotopes. When the water is warmer, more of the lighter isotopes are absorbed and incorporated in the otolith. The ratio of the two isotopes varies directly with the temperature of the water. If we know the isotopic composition of the water as stored in the otolith, we can infer past water temperatures."

Variations in the width of the rings, the scientists said, can also reveal ecological and physiological stresses like starvation, mating or especially cool winter temperatures.

The Michigan researchers have already extended their investigation and identified isotope fluctuations in 3.5-million-year-old otoliths from Florida and Idaho. One of the earliest known otoliths, a 200-million-year-old specimen from England, has yet to be analyzed.

The more immediate objective is to examine stones from fish ears to chart seasonal temperatures in the Great Lakes region over the past 11,000 years. With otoliths as their paleothermometers, the scientists expect to learn enough about past climates to make more meaningful predictions of future climate change.

—JOHN NOBLE WILFORD, December 1992

"Handedness" Seen in Nature, Long Before Hands

Patricia J. Wynne

Fossils of trilobites show characteristic bites that may have been made by the Anomalocaris, an ancient crustacean, shown above with sponges from the Burgess Shale.

BITE SCARS FOUND in the fossils of ancient shellfish have yielded strong evidence that half a billion years ago animals had already evolved "handedness," the kind of behavior that makes most human beings favor one hand over the other.

A new study shows that certain sea creatures that evolved even before animals first colonized the land—and long before the evolution of the hand—persistently turned toward their favored sides when attacking prey or evading predators.

Until about two decades ago scientists generally believed that only human beings exhibited handedness. One reason was that while handedness in human beings is obvious, a propensity to favor one side or another in other species is usually difficult to identify.

But recently asymmetrical behavior has been discovered in many living animal species, and paleontologists have also begun to find hints of preferentially "handed" behavior even in long-extinct animals.

The latest evidence comes from an investigation by Dr. Loren E. Babcock, a paleontologist at Ohio State University, and Dr. Richard A. Robison, a paleontologist at the University of Kansas. The subjects of their inquiry were trilobites, ancient shellfish distantly related to modern horseshoe crabs and crustaceans. Trilobites flourished from about 550 million years ago until they became extinct 230 million years ago, just before the dawn of the age of dinosaurs.

Trilobites are named for their body shape, which consisted of three lobes: a central "axial lobe" containing the creature's essential organs and two side (or "pleural") lobes. Trilobites had many pairs of legs that probably helped propel them along the bottom of the sea, and some trilobites developed large flat tails that may have been used for swimming, in somewhat the way a whale propels itself with its tail flukes. Trilobites had hard shells, the details of which were often very well preserved as fossils in sedimentary rock.

Although they flourished for a very long time, trilobites were food for many predators in the ancient seas.

In his study of hundreds of trilobite fossils, Dr. Babcock was struck by a surprising pattern in the scars left by bites they had sustained. He found that the overwhelming majority of these scars were on the right rear parts of the animals' bodies.

In recent issues of *The Journal of Paleontology* and of *Natural History* magazine, Dr. Babcock reported that the pattern of bites on trilobites' tails strongly suggests that the animals were fleeing predators when attacked. But the preponderance of bites on the right side implies something less obvious: the existence of handedness.

One explanation, Dr. Babcock said, might be that most trilobites themselves favored their right sides, and that they instinctively veered to the right when trying to escape. Another explanation could be that the predators were predominantly "left-handed," tending to attack the trilobite's right rear. A third possibility is that both predators and prey exhibited preferences in their behavior. From the present evidence, there is no way to tell which of these explanations is correct, Dr. Babcock said in an interview, "but there seems no doubt that handedness was common among the creatures of the Cambrian and post-Cambrian periods."

Dr. Babcock examined more than 300 trilobite fossils, all exhibiting malformations of various kinds. Some of the wound scars were attributable to accidents that might have happened while trilobites were in their soft-shell phase, while they were mating or in some other way. These trilobites were discarded from the survey. But of the animals clearly displaying healed bites, more than 70 percent had been injured on their right sides.

Evidence from well-preserved fossils shows that a trilobite's organs were essentially symmetrical, so there is no apparent physical reason why a predator should find one side of a trilobite particularly tasty.

Scientists cannot be sure which species were the main enemies of trilobites, but the shape and size of many of the bites chewed from trilobite shells match the circular, nutcracker mouth of a primitive animal called Anomalocaris—a 20-inch-long creature with no known modern descendants. Fossils of Anomalocaris are abundant in the same sediments as those of the trilobites.

Dr. Babcock said he believed that asymmetric behavior helped species to survive, and that handed preferences have probably been common in animals, plants, bacteria and even fungi almost since the dawn of life.

In many cases, handed behavior is related to nonsymmetrical body parts or some asymmetry in the function of an animal's nervous system. The manual preference of right-handed people, for example, is believed to stem from the dominance of the left hemispheres of their brains. Also, the right hands of right-handed people tend to be slightly larger than their left hands, Dr. Babcock said, and some glove manufacturers make right gloves about 4 percent larger than their mates.

Present-day lobsters and other crustaceans generally have one dominant claw, probably because of some asymmetry in their nervous systems. Lobsters can adapt themselves to favoring the other claw, however, if the dominant claw is injured.

The right and left ears of owls are set at different heights, Dr. Babcock said, and there is evidence that the two ears have somewhat different sensitivities to various frequencies of sound, an attribute that probably gives the owl some acoustic advantage in seeking prey.

"There's probably something asymmetrical in every organism, and it's likely that both physical asymmetry and behavioral handedness confer survival benefits," he said.

The origin of handedness in living creatures is unknown, but one popular hypothesis is that it has a chemical basis.

Chemists have long known that many molecules essential to life exhibit chemical handedness, or "chirality." Amino acids, the building blocks of all living tissue, are molecules that can exist as left- or right-handed twins, each the mirror image of the other. But for some unknown reason, only the left-handed versions of amino acids are created and used by most live organisms.

A growing appreciation of the differences between left- and right-handed enantiomers—molecules that differ from each other only in the handedness of their structures—has caused a revolution in the pharmaceutical industry, which now must worry about the handedness of many of its products; in some pairs of mirror-image molecules, one partner is beneficial while the other is dangerous. The manufacturer must take care to produce only the desirable molecule while excluding its unwanted mirror image, a task that is often difficult and expensive.

Physicists have sought to explain the handedness that pervades life on earth in terms of the subtle interactions between subnuclear particles. With the discovery of the weak nuclear force, which mediates some types of radioactive nuclear decay, scientists found a flaw in the traditional assumption that the forces of nature cannot tell right from left. Weak nuclear interactions, physicists discovered, actually have a slight left-handed bias.

Theorists in the 1980's hypothesized that when weak nuclear interactions impinge on the electromagnetic processes that drive chemical reactions, the weak interactions would favor the formation of left-handed amino acids. Thus, scientists suggested, the fundamental left-handed bias of the universal weak nuclear force might be responsible for handedness in chemistry, biology and many aspects of animal as well as human behavior.

But this appealing theory suffered a blow two years ago when three Russian physicists calculated that the effect of the weak force would be far too small to impart any bias in the handedness of amino acids.

Whatever the origin of handedness, biologists agree that it must be beneficial to the creatures that exhibit it, although it is not always clear why. It is now thought that even left-handedness in human beings, widely regarded as a liability, is not the impediment it was once considered.

In 1989, left-handed people had something of a scare when psychologists at the University of British Columbia concluded from a survey that left-handers were almost twice as likely to suffer serious accidents as were right-handers. But last February, a new and more comprehensive study concluded that no such risk existed. A team from the National Institutes of Health and Harvard University collected data for six years on 3,774 people, and found that the risk of accident was no greater for the left-handed than the right-handed.

"Whether it favors the right or the left, it seems that handed behavior is important to life itself," Dr. Babcock said.

—MALCOLM W. BROWNE, June 1993

2

FRESHWATER

FISH

A 400-million-year gulf divides fish and humans, yet many features of fish biology are easily recognizable. Some species have social hierarchies. Some go in for elaborate mating games. Fish that inhabit the seasonally flooded forests of the Amazon are partial to fruits and nuts, whose trees depend on them for dispersal of their seeds.

The kinship of fish and humans as fellow vertebrates is likely to be become even more evident in future. The fruit fly has long been the creature of choice for biologists seeking to understand how animals are put together at the molecular level. But since flies lack equipment such as hearts and blood and backbones, scientists have recently adopted the zebra fish as a model organism. Biologists who study Drosophila fruit flies, and are aware of how many genes Drosophila and humans have in common, sometimes convince themselves that humans are mostly large, wingless flies. As the zebra fish's genes become known in comparable detail, some biologists will doubtless come to regard humans as finless fish.

Such a view would not be wholly meritless. Fish have chosen to inhabit a different medium from us, but in a world where the numbers favor insects and worms and other alien orders of being, fish are almost kin. It's easy to see some family resemblances, at least, in the following articles, which describe recent research on the biology of freshwater fish.

Guppies Hint How to
Attract a Mate

GENTLEMEN! Ask yourselves the following question. If you were hoping to attract the woman of your fantasies over to your table in a bar, would you rather be sitting next to a guy who looks like Mel Gibson, or like Jerry Lewis in *The Nutty Professor*? Against which backdrop are you likelier to shine, or even appear visible at all? Think carefully and realistically—the future of your chromosomes may depend upon it.

Scientists have devoted thousands of hours studying the details of how female animals choose their mates and how that pickiness drives the evolution of male finery and flamboyance. They have also looked at how males fight one another for access to those females, engaging in ritualistic displays of sound and fury known among biologists as male-male competition.

Now researchers have discovered a new twist to the struts and frets of animal courtship. They have found that a male will preferentially congregate with other males who have proved to be lousy Lotharios, the better to appear himself as a comely alternative—or at least the lesser of two evils.

Studying the behavior of guppies, colorful little fish that have made sizable contributions to evolutionary science, Dr. Lee Alan Dugatkin, of the University of Missouri in Columbia, and Dr. Robert Craig Sargent, of the University of Kentucky in Lexington, have determined that males will observe the performance of their competitors to see which the females prefer and which they avoid. When given a choice to swim beside a loser at love or a winner, the observing males overwhelmingly opt to situate themselves cheek by gill with the chump.

The latest results, which appear in a recent issue of the journal *Behavioral Ecology and Sociobiology,* not only demonstrate that males will go to any length to appear good to females, but that even fish with brains as small as pinheads are capable of surprisingly sophisticated social behavior.

"This kind of behavior requires an if-then sequence of thought," said Dr. Dugatkin. "If I see a particular outcome with another male, then I behave accordingly. It's not the sort of ability you normally expect from a guppy."

Dr. David Sloan Wilson, a professor of biology at the State University of New York at Binghamton, said of the research, "The message that comes out of all these stories is that we have to revise upwards our estimates of the cognitive abilities of nonhuman animals."

Dr. Dugatkin had earlier shown that female fish often make their choice of a mate through imitation, observing which male their peers prefer and then picking the same male. The latest research suggests that males, too, act on their assessments of their peers, though to very different ends.

In the experiments, the scientists rigged up a tank with various partitions to place fish closer together or farther apart. They allowed a male to observe a female swimming in a tank with another male that either was permitted to be close to her or was kept away from her by an invisible partition. The distance between the female and the male under observation served as a controllable proxy for her preference. In general, the closer a female guppy stays to a male, the more she fancies him.

The observing male was then put in a tank with both the faux winner and faux loser in the battle for the female's affections. In 24 out of 30 independent trials with 30 different sets of guppies, the observing male swam right over to the part of the tank where the loser was kept, and lingered there as though he just happened to like his company.

To assure that the observing male was not merely responding to indirect cues from the victorious male, whose recent proximity to a female could in theory have aroused him and made him more aggressive and unapproachable, the scientists also did the same experiments without letting male No. 1 observe the interactions between the two males and the female. When the male had no clue who had been beside the female and who had been kept away, he chose to swim beside one or another with equal frequency.

Dr. Dugatkin admits that the experiment was performed under artificial laboratory conditions. He hopes soon to observe the guppies in their native Trinidad to see if they perform similarly in the wild. Nevertheless, he proposes the results are meaningful in their surprising consistency. Among guppies, which aggregate in loose schools of 15 to 100 fish, males cannot monopolize territories or harems and therefore have trouble making themselves stand out to potential mates. Finding a less attractive foil to underscore one's graces may be a male's easiest route to reproductive success. Dr. Dugatkin also suggests that such a strategy of conquest by contrast is not likely to be limited to guppies, but may apply to other social animals like birds, dolphins or primates, in which individuals are perpetually assessing the strengths and weaknesses of their neighbors.

"I wouldn't hesitate to say the same things may be going on with humans," Dr. Dugatkin said. He is hoping to do experiments shortly to see if people, like guppies, choose their cruising companions in part to make themselves appear the choicest catch of the sea.

—NATALIE ANGIER, November 1994

They're Smart, for Fish, and a Model of Diversity

THE DATE IS A DUD and both parties know it. Yet as long as they are stuck with each other for a time they make a wan effort to flirt. He lunges lazily toward her. She quivers gently in response. He flaps his tail against her. She flares her gills to show their provocative red undersides. He circles around, charges her again and tries to nip her, but now she's getting bored with the charade and moves away from him. Reacting likewise, he drifts off to the opposite end of the tank. For a few moments they are each lost in the inscrutable vastness of fish thought. And then it happens. The female opens her plump, sensuously carved lips into the widest, roundest, most perfect, least courteous gape of mouth that can be imagined: a fish yawn.

"The female doesn't seem very interested, does she?" said Suzanne Henson, a student carrying out an experiment on the mating habits of cichlid fish. "She's not doing the sort of things you'd expect from a receptive female. She's not doing the slip motion, gliding her whole body along the body of the male."

Sometimes, Ms. Henson said, when a female is put in a tank with a male, she becomes so excited that her genitals swell and she immediately grows heavy with eggs. For his part, an interested male is a violent male, behaving toward his potential mate with an abusiveness that looks like grounds for criminal charges. "Once a male bit a female so hard I actually jumped," said Ms. Henson. "I could hear the sound of the crunch." But not today, and not with these two slugs. Their disastrous date is finished, the experiment over, and each is returned to its proper tank.

Ms. Henson works in the laboratory of Dr. George W. Barlow of the University of California at Berkeley, a leading authority on the great and wildly diverse group of animals known as cichlid fish. She and others in the lab are studying the Midas cichlid, a beefy, square-jawed creature from Nicaragua that comes in two color schemes, zebra-striped or gold—the last accounting for the species' name. Midas fish, like many other cichlids, are monogamous, and the researchers are seeking to understand the individual traits that inspire one Midas to choose another as its mate for life.

The question is part of a broader consideration of the sexual, social and feeding behaviors of cichlids, an extraordinary family of fish that many evolutionary biologists believe could help resolve the great puzzle of how species evolve and how diversity in nature arises from monotony. Assisting the new behavioral studies are molecular analyses of cichlid DNA, through which scientists are trying to determine relatedness between species and to map out the many twigs on the cichlid's dense and tangled family tree.

More than 1,000 species of cichlid fish live in the lakes and rivers of Africa, Madagascar, India and Latin America. They are a highly successful tribe, frequently dominating their environment through a blend of intelligence—unusually high for a fish—and elaborate rituals of parental care. But what makes them so unusual is the number of species that often coexist in the same place. Over 500 different varieties of cichlids swim in Lake Malawi, in southeast Africa, while about 200 other species live in Lake Tanganyika, in Tanzania. Some species are bigger than goats, others could fit in a thimble. Some are thick and boxy, others lean and long. They are brown or turquoise or every shade of a neon rainbow painted on a single beast.

And the cichlid's rate of speciation has been explosive. In Lake Victoria of East Africa, for example, 300 species of cichlids arose in less than 200,000 years, maybe even as little as 12,000 years, an evolutionary pace that no other animal group has rivaled. Certainly none of the other fish groups found in the three African lakes has undergone anything approaching the spectacular diversification managed by the cichlid family.

Scientists have long been captivated by cichlids, seeing in them a far greater opportunity to probe essential evolutionary patterns than was afforded by another famous family, Darwin's finches. Until recently, much of the research has relied on traditional taxonomic and observational approaches, tallying up species by studying fish anatomy, as well as by watching fish behavior. Now biologists have added molecular analysis to their research, tracing cichlid lineages and cichlid radiations by studying the fish's DNA.

In a recent issue of the journal *Trends in Ecology and Evolution,* Dr. Axel Meyer, a molecular geneticist at the State University of New York at Stony Brook, pulled together much of the recent molecular data on cichlid fish. The DNA work has confirmed previous results from the taxonomists that cichlids are monophyletic, that is, they all originate from a single ancestral fish that arose perhaps 120 million years ago, when India, Africa and Latin America were one giant continent. Since the breakup of the continents, the founder fish that were carried off to different regions of the planet have gone their own ways, speciating wildly in all cases yet by very distinctive genetic mechanisms from one lake or river to another.

In some instances, species of cichlid fish that look and behave radically differently from one another turn out to be almost identical genetically.

For example, Dr. Meyer compared the DNA of 14 Lake Victoria cichlid species, choosing fish with radically divergent feeding behaviors: a snail eater, a cichlid that feeds on its fellow cichlids, a cichlid that eats only the eyes of other cichlids, another that exclusively sucks young cichlid fry out of the protective mouths of their parents. Yet despite the fishes' specialized appetites, their genes differ from one another by a mere two or three bases, or chemical subunits, out of the many hundreds of bases that constitute the genes examined. "This genetic invariance was a very big surprise to us," said Dr. Meyer. "There's more variation among human populations than I had among my fish." And humans, of course, are all members of the same species.

The new work suggests that much of the success of the cichlid family could lie in its unusual degree of molecular flexibility, with minor differences in genes able to yield enormous disparities of comportment. And it is the cichlid's ability to specialize, scientists believe, that helps explain how so many species can live cheek by gill in the same body of water with each still managing to earn a living. If all cichlids were bottom grazers, for example, one species would likely outcompete the others into oblivion. But each cichlid has evolved its own hunting method, and each strategy seems more bizarre than the last. There is a cichlid that resembles a rotting fish and spends a lot of time floating as though dead; but when another fish approaches, thinking it has happened on an easy meal, the corpse springs to life and attacks the would-be scavenger.

A recent paper in the journal *Science* describes a newly discovered cichlid in Lake Tanganyika that has its head bent permanently to the left, an adaptation that enables its teeth to efficiently scrape a meal of scales off the right side of a passing fish's body. More improbable still, the researchers found a second type of cichlid that has evolved a head curving to the right, the better to shave scales from a prey fish's port side.

"There's always a new amazing story when you study cichlids," said Dr. Meyer. "The standard idea in ecology is that there are various niches waiting to be filled, and species arise to fill them. But cichlids seem to create their own niches."

Scientists believe that in a lake like Victoria or Tanganyika, the cichlids that originally founded the flocks were generalists, which then became specialists as competitive pressure increased.

Some scientists have suggested that cichlids have been able to evolve so many eating strategies by the grace of an unusual feature: They have two sets of jaws, one in their mouth, as the average fish does, and a second in their throat. With the throat jaws available to process food, the mouth jaws can be extremely flexible, evolving very specific methods for capturing food. "The idea here is that if you split the function into two jaws, there's less evolutionary constraint," said Dr. Melanie Stiassny, a cichlid expert at the American Museum of Natural History in New York. "In essence what you have here is a throat jaw that's a jack of all trades, and a mouth jaw that's a master of one."

Variations in dining strategies, however, are not always the major distinguishing traits of cichlids. Reporting in the journal *Nature* last year, Dr. Meyer and Dr. Christian Sturmbauer, a co-worker, looked at the DNA of six cichlid species from Lake Tanganyika and found considerably more genetic variation than they had observed in Lake Victoria cichlids, but in this case the predominant differences among the species were in their colors. And many cichlid biologists now believe that coloration holds another essential key to the fish family's story, for coloration often goes with sexuality and mating preferences, among the more potent driving forces in evolution.

Cichlids have always been popular fish among home aquarium hobbyists; Wanda, in the movie *A Fish Called Wanda,* was a South American cichlid. Fish keepers claim cichlids are so bright, they recognize individual humans, but people are especially taken with the fish's courtship and fry-rearing practices. Most fish lay eggs and abandon them, or the father may remain to watch the eggs until they hatch. But among cichlids, both parents often engage in protracted parental care. They brood their eggs in their mouths, and even after the fry are born, they protect the little fish by taking them back into the safety of their mouths when predators approach. "They'll suck the fry back in as though they're sucking in strands of spaghetti," said Dr. Barlow of the University of California at Berkeley.

The habit of mouth brooding has led to a few outstanding features on male cichlids. Because predatory pressure in a cichlid's habitat can be extreme, many females, after laying their eggs, frantically turn around and begin scooping them into their mouths before the eggs have been fertilized. Males have adapted to this by evolving bright spots on their rear fins that strongly resemble eggs. When the female is sucking in her eggs, the male gives his rear fin a shake, she tries to take the dummy eggs into her mouth, and—*whoosh!*—the male releases a stream of semen into the female's mouth that then fertilizes the eggs.

In some species, both parents also feed their fry with their own flesh, allowing the young fish to nibble at the scales and nutritious mucus cells on the surface of their bodies. "The parent is a big breast is what it amounts to," said Dr. Barlow.

Given the high investment that parents make in their young and in each other, scientists propose, cichlids must have ways of selecting worthy part-ners. Fish are visually oriented, so it is likely that they pick mates based on cues of color. Dr. Meyer suggests that sexually selected traits like color may undergo far more rapid divergence than would traits that affect an animal's ability to survive, and hence could partly explain the explosive evolution of cichlids.

In experiments with the Midas cichlids, Dr. Barlow and his co-work-ers are trying to tease out the details of mate preference. They have learned that although only about 8 percent of the species develop a gold coat, both other golds and the dull zebra-stripers prefer a mate of gold when given the choice.

That could be because the golds look more threatening. Cichlids must often fight off outsiders when rearing their brood, and so toughness in a mate is highly valued. Through their detailed matchmaking trials, the sci-entists have learned that mate choice proceeds in two steps. First, the female finds a male who appeals to her, for reasons that the scientists have yet to glean. But once the female has demonstrated a liking for the male, he will start exerting his own choosiness by behaving extremely aggressively toward her.

"He's testing the female to see if she's aggressive enough," said Dr. Barlow. "She's got to threaten him back in the right way if he's going to accept her." Once the male has determined that the female is tough enough, he will mate with her and treat her gently ever thereafter.

The odds of a male and female cichlid sharing just the right color and chemistry are slim, which is why so many Midas encounters end in fish ennui and a giant fish yawn.

—NATALIE ANGIER, August 1993

In the Amazon's Depths, a Strange New World of Unknown Fish

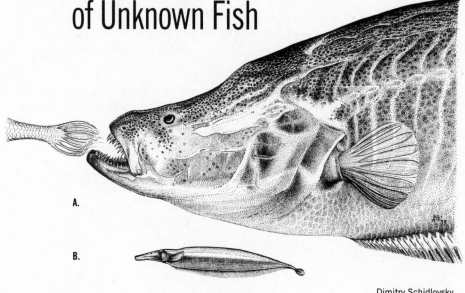

A.

B.

Dimitry Schidlovsky

C.

D.

A. *Magosternachus duccis*
This predatory electric fish eats chiefly the tails of other electric fishes, which then regenerate.
B. *Orthosternachus tamandua*
A nearly blind electric fish with very small eyes from deep in the Amazon, formerly known from only two specimens preserved in museums.
C. *Bathycetopsis ollvelral*
A miniature, blind and pigmentless catfish from the Amazon's main course. It grows no longer than about two inches; its closest relatives are three to four times that size. Not only is it electroreceptive, it has taste buds all over its body.
D. *Magosternachus raptor*
Another previously unknown species of predatory electric fish. They are able to hunt and navigate without vision in the muddy, nearly lightless depths by generating sensitive electric fields around their bodies.

SINCE THE MIDDLE OF THE LAST CENTURY, the Amazon River and its tributaries have provided passage to many an intrepid adventurer heading into South America's interior rain forests. But despite the fact that scientifically minded explorers have been floating down these waterways for more than 100 years, life in the river itself has remained largely unknown.

A handful of scientists have now begun the difficult work of plumbing the Amazon's depths, casting down their nets to unveil the deep-water world below. With each haul, long-hidden inhabitants of the Amazon, the largest tropical river system on earth, are coming to light, including oddities like transparent catfish and electric fish that subsist solely on the tails of other electric fish.

"We've now gone close to 2,500 miles over the Amazon and its tributaries,"said Dr. John G. Lundberg, an ichthyologist, or fish specialist, at the University of Arizona in Tucson. "You come up with drastically different kinds of fish."

Dr. Richard Robins, the Maytag professor emeritus of ichthyology at the University of Miami, said, "Lundberg's work has been pioneering." Dr. Robins helped make the case to persuade the National Science Foundation to finance Dr. Lundberg's team when it first proposed fishing the deep rivers. He argued that even though the team would probably lose every net it had trying to catch fish from the ragged bottom of these rushing rivers, if it got just one successful haul, the effort would be worthwhile.

"No one knew what was going on down there," Dr. Robins said. "It's a big breakthrough."

Like the Amazon rain forest, the Amazon River is home to creatures of extreme diversity, helping to make South America the continent with more fish species than any other in the world. So far, the Amazon and its tributaries are estimated to harbor at least 2,000 freshwater fish species, twice the number in the United States, Canada and Mexico.

"It's just an overwhelming diversity," said Dr. William L. Fink, an ichthyologist at the University of Michigan's Museum of Zoology, "and how all this came to be is a real interesting question. The deep-water faunas are especially odd and the most challenging to get to. You get into strange and interesting new worlds."

So far, Dr. Lundberg and his collaborators, including Dr. Cristina Cox Fernandes of Brazil's National Institute of Research of Amazonia and Dr.

Naercio Menezes of Brazil's Zoology Museum in São Paulo, have amassed 125,000 fish, and their ever-rising species count is at 240. What they have uncovered at the greatest depths is a murky, nearly lightless world inhabited by a proliferation of two kinds of creatures: electric fish and catfish.

"The Amazon water is muddy," Dr. Fink said. "It's hard to see just a few feet down, and 20 feet down, it's really black." Researchers say that may explain why two sorts of fish ready-made for life without light lurk at the bottom.

Electric fish are able to hunt and navigate without vision, using electric organs to generate electric fields around their bodies to sense where things are. Catfish are also electroreceptive, and they also have taste buds all over their bodies, allowing other senses to dominate over sight.

The most peculiar among the electric fish found by Dr. Lundberg and his colleagues are two species of tail-eaters. When researchers examined the fish, their stomachs were filled entirely with the tails of other electric fish. Dr. Lundberg said that while the researchers know the fish eat the tails of other species, they may eat the tails of members of their own species as well. The electric fish can rapidly regenerate lost parts, which makes the meal of choice of these species both plentiful and naturally renewable.

Among the electric fish, the scientists also discovered two species that are so similar the researchers were not sure that the fish were distinct species until they saw their electric organs discharge. John P. Sullivan, a graduate student at Duke University, discovered that these fish, which appeared to be mottled and solid-colored variants of a single species, were in fact two distinct species, each with its own unique blips of electrical discharges.

Dr. Lundberg and his colleagues also found many eyeless or nearly eyeless catfish and electric fish. Among the blind fish, researchers netted a tiny transparent catfish that was bearing eggs when it was just one third of an inch long—close to the world record for the smallest fish at sexual maturity. Though tiny, blind and transparent, this species may not be quite so vulnerable as it sounds. The fish have greatly thickened bones and armor plates on their sides.

The researchers have found some novel features on some fish that they cannot explain. An electric fish found in only one small area of the Río Negro, a tributary of the Amazon, has a unique tongue-like projection just above its chin.

"It's not a tooth—it's a soft organ," Dr. Lundberg said, adding that nothing like it had been seen on any electric fish before.

As predicted, the team has lost a lot of equipment. Motoring along the Amazon, researchers drop a weighted trawl net that has a wide mouth and a mesh fine enough to capture the tiniest fish.

These nets sink out of sight into water that is racing by at speeds of six feet or more per second. After towing the net, researchers must carefully retrieve it from depths of 30 to 150 feet below the surface. But they must avoid tearing the nets on the debris-covered bottom, spiked with dead trees, or on the water-logged branches that can be racing by at any depth in the river. Should any equipment come loose, it is gone, as the current sweeps everything quickly away. And there is no way to jump into these waters to work. "That far from shore, you'd get swept away downstream," Dr. Lundberg said. "There's no way a person could withstand that."

But it is not just rushing waters that have kept the fish in the Amazon and its tributaries mysterious. One problem is the sheer size of the Amazon basin, which stretches over some 2.5 million square miles of equatorial South America. Moreover, some stretches of river are very remote. And as Dr. Richard Vari, a research zoologist at the National Museum of Natural History at the Smithsonian Institution, explains, scientist-explorers are not always welcome. "In Peru and Ecuador, and Colombia in particular," he said, "people think this must be some sort of cover for spying on the drug traffic. They find it very difficult to believe that you're really out there in these remote areas just to collect small fish."

As a result, even many intriguing and easily caught species remain poorly known. For example, there is a small group of fish that lives along the banks of the Amazon that actually eats driftwood. Dr. Scott Schaefer, associate curator of fishes at the American Museum of Natural History in New York, said those fish are the only vertebrates known to live by eating dead wood, yet "we know nothing about their biology."

The best-known species are those with commercial value: the large fish for market and the small, colorful fish living in the river's margins that are important in the aquarium trade. Yet even for such commercially valuable species, the most basic information is often lacking.

Dr. Lundberg and his colleagues found that the young forms of many commercially important catfish were not living in shallow habitats, as some had thought, but were swimming in the main water column of the river. It is just this sort of information that is crucial for managing the species that are threatened with overfishing.

As the researchers work on, their findings are available on the Internet. Dr. Lundberg maintains an Internet site on the collecting expedition: http://eebweb.arizona.edu/fish/calhamaz.html. Dr. Fink is in charge of a World Wide Web site sponsored by the National Science Foundation with data from around the world on collections of fish from Central and South America: http://www.keil.ukans.edu/neodat/.

There is one point in the Amazon—a huge, gaping hole near the mouth of the Río Negro—where the river bottom drops more than 300 feet. Dr. Lundberg's response to the suggestion that this might hold the most interesting fish?

"I'm not going to put a net down in that thing," he said. That abyss is out of even his depth, leaving the Amazon, for the moment, to keep at least one last mystery to itself.

—CAROL KAESUK YOON, February 1997

In the Quiet World of Fruit-Eating Fish, a Biologist Feels Too Alone

IN BELÉM, BRAZIL, Michael Goulding works in what he calmly describes as "the greatest evolutionary theater in the world." One of the few scholars to study the rivers of the Amazon basin, he spends months sifting through muddy streams in a region where a single river holds more species than all the waters of North America.

The 38-year-old California biologist can barely hide his joy when he talks of recording more than 400 fish species that were never before identified. He dwells with enthusiasm on the unusual relationship between fish and trees he found here in the Amazon, a discovery that made him famous among fellow biologists.

A fieldworker by temperament as well as philosophy, he sometimes eats peccary for lunch and curassow, a turkey-like bird, for dinner. His neighbors include electric eels and poisonous ants and the spot where he sleeps may suddenly cave in and slide away.

Now, at a time of escalating concern about the cutting of tropical forests, a time when governments and international aid agencies are joining conservationists in praise of the biological riches of the tropics, one would expect someone like Dr. Goulding to feel pleased. Instead, he is alarmed. For even as concern about the fate of tropical ecosystems seems to rise, he notes, the study of tropical nature in the field has been drained of resources, prestige and qualified people in favor of highly technical laboratory work.

Like many other experts in tropical ecology, Dr. Goulding warns that, with tropical forests being destroyed and rivers permanently altered at a rapid rate, many species are becoming extinct even before they are discovered.

"Natural history has become unfashionable," said Dr. Goulding. "It cannot compete anymore with looking at cells and splicing genes." Cultural perceptions of what is "scientific" have changed, he said, and few biologists are motivated to continue the fieldwork of natural history.

"Some people come here on expensive projects to study the biochemistry of frog saliva or the brain cells of a certain fish," he continued. "What good does that do here if we don't even know most of the species and organisms?"

As he walked along a bank of the mighty Para River, Dr. Goulding gestured at the vast expanse of water before him. "The Amazon basin is so huge, we don't even know the generalities," he said. But the answer is not to start "hastily making inventories as some institutions want."

"Making up lists of species is not that vital," he argued. "We need to understand the interactions, how the system works, so we can learn from it."

Achieving such an understanding requires long, uncomfortable and often lonely periods of observation. "Laboratory work pays more and you don't run risks like getting malaria or hepatitis," said Dr. Goulding, who maintains a cluttered office in Belém but spends much of his time on a boat in an Amazon tributary or camped on a riverbank. Sometimes he has to wait an entire day for tropical rainstorms to pass. All sorts of creatures in the water, in the air and on land can deliver poisonous bites.

But the rewards are staggering, he said. For example, the Río Negro, the huge tributary of the Amazon, has habitats as diverse and rich as that of a coral reef, he said. "The Río Negro alone has some 700 species," he said. "The U.S. and Canada together have 500."

Upriver, he uses dugout canoes to wind his way up inlets and visits Indians and river people. "The Indians have great knowledge of natural history," said Dr. Goulding. "They talk all the time about plants and animals. That knowledge is disappearing as more and more Indians assimilate."

To illustrate the importance of the study of interaction, he cites the rich life cycle set in motion every summer when the cresting rivers flood about 40,000 square miles of forest. Though the phenomenon was long known to the Indians, Dr. Goulding was the first scientist to document fully how fish live off the forest, eating leaves, insects, fruits and seeds. Some, like the piranha-caju, even have large molars and strong jaw muscles to crack nuts.

The larger fish species in turn help reproduction of trees because they eat, transport and excrete the seeds.

One perfectly adapted fish is the tambaqui, a deft cousin of the piranha, which can grow up to three feet long and weigh 60 pounds. It has taken on camouflage colors, black and olive or moss green, hard to see in the dark flooded forest. But its most impressive feature, Dr. Goulding said, is its set of teeth. It has developed broad molars to crush seeds and nuts and it has incisor teeth as well as long and fine gill-rakers that the younger fish use to capture animal plankton.

He found that half of the 96 tambaqui he examined in one study had crushed rubber tree seeds in their stomachs and one fifth contained masticated palm fruits, which can be up to two inches in diameter.

Another biologist, Ivan Sazima, who studied other fruit-eating fish in Western Brazil, said the fish either wait for the fruit to drop or pick them directly off branches that hang into the water. "The fishermen imitate the trees," he said. "They know that if they drop a fruit vertically into the water, they can attract the pacu."

But as the flood plains are increasingly deforested by cattle ranchers and farmers growing jute and rice, the damage is evident, said Dr. Goulding, whose research has been financed by the World Wildlife Fund, in part to study management of fish for food. "The forest is cut, there are less places to feed, and the fish available for food drops," he said.

The Amazon may have the world's largest freshwater fish: the pirarucu grows up to 15 feet long. "But most fishes here are only one or two inches," he added.

Many biologists say that in the tropics, it is even more urgent to study the river systems than the vegetation because river life is facing more radical change. In Amazonia, the growing cities and industries, deforestation and new dams are altering the chemistry of the rivers over thousands of miles.

"Dams have changed the ecology of some rivers forever without anyone ever having studied their natural history," said Dr. Goulding. "In cataracts and rapids, there are whole groups of fish that are adapted to turbulent waters. Some are completely blind and they have strange large fins. Then suddenly everything changes when a dam is put in."

But it is unlikely that the world's greatest freshwater system can be adequately studied before it is irrevocably changed, he said. "The task is so great and we are so few people studying the natural history and the ecology," he said. "It cannot be finished in this century."

Dr. Goulding is not alone in voicing his concern over the state of tropical field studies. A report of the National Research Council in Washington, an arm of the National Academy of Sciences, has urged governments to give high priority to tropical biology "while there is still time." The report called for a "fivefold increase" of systematists, the researchers who classify organisms and the relationships among them. There are no more than 1,500 systematists trained to deal with tropical organisms. Their number is getting smaller, the report said, as professional opportunities have declined and priority has been given to other disciplines.

Peter Raven, a biologist who directs the Missouri Botanical Garden in St. Louis, said: "There is almost no information on most groups of organisms, and all the information is slipping through our fingers because we are passive about it."

The tropical countries have historically spent little on biology, Dr. Raven said in an interview. The United States National Science Foundation earmarked only $4 million for "tropical systematics" in a budget of $1.6 billion for the fiscal year 1987.

Europe, including Britain, which produced many of the great naturalists of the 19th century, has also, by all indication, lost interest. "The Europeans are backing off even faster than the Americans," Dr. Raven said.

—MARLISE SIMONS, February 1988

For the Male Molly,
It's Not So Sterile an Affair

IT HAD LOOKED TO ALL THE WORLD like a sterile affair, one built on mutual convenience and exploitation. She, an Amazon molly fish, had sought copulation with a male, not to obtain his genes and mix them with her own to spawn the next generation, but simply for a quick chemical jump start. She needed his sperm to prod her eggs to split in twain and create clonal progeny, the fry all female and identical to the mother.

For his part, the male, a closely related species of molly fish that propagates through more normal channels of sexual reproduction, also appeared to benefit from the odd affair. As scientists showed last year, the male molly fish that copulates with an Amazon becomes immediately irresistible to the females of his own kind, who then hungrily mate with him and give his seed a chance to sire piscine multitudes. In this way, scientists proposed, a so-called gynogenetic species like the Amazon can persist, gaining the chemical prod of a male's sperm without giving him offspring in return. She merely made him sexy to other females, and that seemed motivation enough.

Now, however, researchers have demonstrated that the union between Amazon and male molly may not be so genetically one-sided after all, and that the male gets more from the dalliance than a buff to his image. Reporting recently in the journal *Nature,* Dr. Manfred Schartl, of the University of Wurzburg in Germany, and six colleagues have found that on occasion, a tiny bit of the male's DNA is transferred to the eggs of an Amazon fish.

The amount that slips in is far less than what accompanies the usual meeting of sperm and egg, when each contributes half the genes to the union. The Amazon fry do end up as nearly perfect clones of the mother. However, during a small subset of matings, pieces of the male's DNA get injected into the egg as so-called microchromosomes, little circles of genetic

43

material set apart from the fish's normal complement of 48 chromosomes. And these microchromosomal structures behave as normal genes, bestowing a few trace traits on the Amazon offspring.

The finding could help explain why gynogenetic species like the fish, some salamanders and a handful of other creatures persist over evolutionary time, rather than rapidly hitting a genetic dead end, as conventional calculations predict they should.

The scientists discovered the microchromosomal contributions by studying laboratory strains of Amazon mollies and the accommodating males of a closely related species, the black molly. The scientists saw that even though the Amazons are normally a light gray color, they occasionally give birth to fry with big, jet-black blotches scattered across their silver scales. Upon screening the chromosomal makeup of the splotchy offspring, the scientists detected unmistakable evidence of genetic material inherited from the black molly father, who supposedly had donated nothing beyond a seminal trigger. Those offspring lacking the black pigmentation likewise lacked any male molly DNA.

The researchers have also shown that in populations of wild molly fish, male DNA on occasion lends an Amazon molly a polka-dot aspect. In theory, genes other than those for coloration are sporadically transmitted from male to Amazon.

Again, both partners of the arrangement benefit. The Amazon molly gets the infusion of new genetic material, which could help prevent her DNA from growing stale or becoming dangerously overloaded with mutations; at the same time, she begets mostly little replicas of herself, which is, after all, what being gynogenetic is all about.

As for the male, he may not donate much DNA to the effort—at best, one hundredth of 1 percent of his genes manage to storm the gynogenetic egg—but those segments that breach the barrier have a chance at near immortality.

"Due to the gynogenetic mode of inheritance, once a piece is in, it's in for a long time," said Dr. Schartl. "It gets transmitted through all future generations, while in normal paternal transmission, a male's genes are diluted down within a few generations." True love may often fade away, but a fleeting affair, it seems, can last forever.

—Natalie Angier, January 1995

In Fish, Social Status Goes Right to the Brain

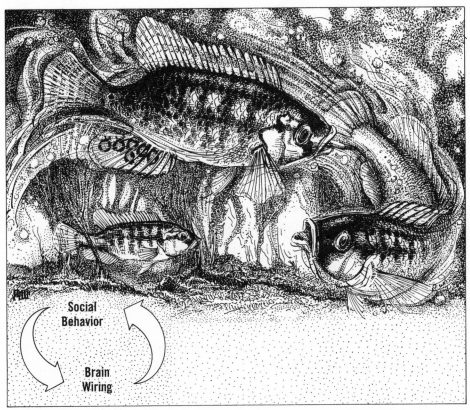

Dimitry Schidlovsky

In cichlid fish, males battle each other for breeding territory. Dominant males, which gain special colors, have now been found to have distinctly larger cells in the brain region known as the hypothalamus. If these males are defeated, the cells shrink and the fish lose their colors and breeding behavior.

AMONG MALES OF MANY SPECIES, from elephant seals to human beings, the struggle to prevail over competing males often seems to dwarf all other tasks—and perhaps with good reason.

A scientist has discovered that how a male fish interacts with other males, and whether it is socially dominant or a tremulous wimp, has such a profound effect on the creature that it changes the brain cells in charge of the fish's capacity to breed.

Studying the African cichlid fish, Dr. Russell Fernald, a neurobiologist at Stanford University, has discovered that in aggressive males that command large territories and keep contending males at bay, brain cells in the hypothalamus that allow the fish to mate are six to eight times larger than equivalent cells in mild-mannered males with no social clout.

What is more, Dr. Fernald has found that the dimensions of those cells are extremely plastic. Should the domineering male be confronted by a larger male able to bully it, the neurons of the defeated fish will rapidly shrink. And after the hypothalamic cells have shrunk, the male's testes follow suit, eventually robbing the fish of its desire and ability to breed.

The discovery is the first persuasive evidence of how social behavior can help sculpt the structure of the brain, and how the altered brain, in turn, influences animal behavior.

"It's a big jump conceptually to ask how behavior might affect cellular and molecular events," said Dr. Fernald, "but now we can look at it directly."

In laboratory experiments that change the social status of the fish, he said, "we can push the males forward or backward, from macho to wimp and back again." Some of the findings appeared in a recent issue of *The Journal of Molecular Endocrinology*.

The results provide powerful evidence that the old nature-nurture argument, pitting the influence of biology on behavior against that of the environment, is far too simplistic.

Instead, researchers said, an animal's body, its behavior and its social milieu very likely represent interdependent points on a giant feedback loop, with one alternately affecting the other.

"This is a marvelous example of how the social environment influences animal biology," said Dr. Darcy B. Kelley, a biologist at Columbia University who studies frog mating songs. "Here you have this animal whose social state is translated into changes in the cells of the hypothalamus, which then contribute to the animal's augmented reproductive capacity. This goes a long way toward solving the nature-nurture debate, which was a false debate to begin with."

The findings may also have implications for a few finless creatures, including people. Because the brain molecules under study are highly conserved across the evolutionary spectrum, Dr. Fernald suggests that the architecture of the human brain may also be affected by a person's behavior.

Indeed, Dr. Simon LeVay of the Salk Institute in La Jolla, California, who stirred an acrimonious debate recently when he said he had detected a difference in the hypothalamus between homosexual and heterosexual men, has expressed great interest in Dr. Fernald's new study.

But researchers concur that it will be far more difficult to decipher the meaning of brain differences in people than it is to investigate the neurobiology of fish.

Dr. Fernald began his research at Lake Tanganyika in Africa, home to more than 150 species of cichlid fish. Studying the species *Haplochromis burtoni,* he noticed that only 10 percent of the males controlled all the feeding territory and all access to females. Those few males, he realized, differed spectacularly from their timid counterparts. Dominant males are larger and more brilliantly colored, with bold orange stripes and fins that gleam like rainbows, and they are extremely aggressive. By contrast, both females and subordinate males have cryptic, sand-colored scales and are unlikely to pick a fight.

But it turns out that the flamboyance of machismo has its price. The colorful fish are easily seen and predated upon, and once a dominant male disappears, a remarkable event occurs. All the meek males in the neighborhood rush over, seeking to fill the vacancy, and a series of violent battles commences.

"They're surprisingly vigorous fighters," said Dr. Fernald. "They chase each other, they bite, they hit each other with their tails. It really looks painful."

Eventually, a winner emerges and promptly begins flaunting his success through telltale displays of dominance and territoriality. At that point a welter of physical changes begins. The male grows bigger and gains a bright coat; its gonads swell and it starts making sperm.

Taking the fish back to his laboratory, Dr. Fernald traced the molecular sequence of events from newfound social dominance to full-fledged sexual prowess. He found that the behavioral changes occur first, and that they in turn spur dramatic growth in brain cells responsible for producing a compound called gonadotropin-releasing hormone. That substance tweaks the pituitary gland to produce hormones that in turn stimulate the fish's testes and switch on sperm production.

Conversely, in experiments where a dominant male is stripped of its pre-eminence by the introduction of a bigger, nastier male, the chastened fish stops making dominant displays and slinks off, and the brain changes follow. Within days, its bright colors have disappeared and its testicles have withered.

"The behavior is what's driving it," said Dr. Fernald. "The failure to play a socially dominant role is the key, which ultimately down-regulates testes size."

That may seem a sad fate for a male who has savored the spoils of power. But then again, said Dr. Fernald, the fish regain their cryptic coloring and hence a potentially longer lease on life.

—NATALIE ANGIER, November 1991

Biologists Embrace
the Zebra Fish

IN PET STORE PARLANCE, it is called a beginner fish: even a fumbling neophyte hobbyist has trouble killing it.

But for a burgeoning group of basic biologists, the zebra fish is an elegant sophisticate, a grand alternative to the flies, worms, mice and frogs they have been studying for decades.

The researchers are realizing that the little striped fish, the very type found in dentists' offices and home aquariums everywhere, offers an advanced means for delving into the mysteries of how a fertilized egg flowers into a complex animal and how the brain sprouts its dense tapestry of interconnected tendrils and synapses.

At first blush the fish may not look like biology's next great hope. An adult is only about two inches long, its colors are drab, and, with its stripes running horizontally from head to fin, rather than up and over its spine, it hardly deserves to be named after a zebra.

But for developmental biologists and geneticists, it is the animal they have long been dreaming of: a small, easily manipulated organism that can be studied in fine molecular detail—just like that staple of the laboratory workbench, the fruit fly—but one that is a vertebrate and thus has all the features of an intricate being like ourselves.

So compelling is the animal that scientists who made their fame by deciphering the genes of viruses, bacteria, nematodes and other creatures are abandoning their chosen organisms for the chancier, but more potentially revealing world of fish. They realize it will take an unforeseen length of time to master fish genetics, and that as they grope about, their productivity is likely to falter.

"I was having a midlife crisis, and I wanted to do something different," said Dr. Nancy Hopkins of the Massachusetts Institute of Technology in Cambridge, who until recently studied retroviruses. "When I first looked into fish, and realized what could be done with them, I fell in love."

The romantic spirit is contagious. Fish aficionados wax passionately about the technical and esthetic advantages of their subject. As embryos, the developing fish are transparent, and thus can be examined under a microscope step by step.

A researcher can watch the entire theater of life unfold, as the fertilized egg cleaves in half and in half and in half again; as the cells glide and shudder into their proper position, forming the black of the eyes and the bulge of the brain; as the pinprick of a heart begins to throb and the wispy veins flush red with blood; as the fin starts to wiggle and the fetus clearly declares itself a fish. And, of critical importance to geneticists, the entire transformation, from fertilized egg to fully formed fish, takes only 24 hours.

As the embryo develops, it can be manipulated with extraordinary precision. Single cells can be picked up with microscopic tweezers and moved to a different part of the embryo to see how they react in their new setting, a type of transplant operation that can be done with no other vertebrate. Individual genes can be mutated to see how the disruption affects growth and behavior.

Fish studies are also very neat. Fetal mice grow in a mother's belly, and thus for any insight into their development, many mother mice must be killed and the messy embryos plucked out. But zebra fish grow in eggs outside the mother's body, where they can be observed without the need for continual slaughter.

Like proselytes, scientists from here and abroad have been flocking to the University of Oregon in Eugene, considered the world's premier zebrafish laboratory, as they seek guidance on how they might open a fish facility of their own.

Dr. Christiane Nusslein-Volhard of the Max Planck Institute for Developmental Biology in Tübingen, Germany, one of the most prominent researchers in the field of fruit-fly genetics, recently persuaded her institution to build a $4 million facility to be devoted to rearing and analyzing tens of thousands of fish.

"Flies are nice, they're wonderful, and it's not easy to stop working with them," she said. "But they're not vertebrates. Their body plan is different, they have no blood circulation, no central nervous system and they have different types of eyes.

"If you want to understand vertebrates, you have to study vertebrates," she said. Besides, she said, "it's great fun to start something new. It's just like playing with toys."

But she and others who are switching to fish admit they are taking an enormous professional risk. The establishment of any organism as a useful model for the big questions of how animals grow and how the brain thinks demands years if not decades of effort, as well as the cooperative contributions of thousands of scientists. Fruit-fly genetics began early this century and has only recently yielded detailed secrets of development. The history of mouse studies is likewise long, giving mouse geneticists ample knowledge to draw from.

But the zebra-fish business is only a few years old, and it is just now gathering a critical scientific mass.

Many technical difficulties remain to be mastered, and an understanding of some elements of fish heritage, such as where different genes are found on the creature's chromosomes, lag behind that for such other intensively studied animals as rodents and humans. Indeed, biologists are still struggling with the most mundane questions, like what sort of food the fish prefer and how to keep the fish tanks clean without spending the entire workday as overeducated janitors. Most worrisome, there is always the chance that the zebra fish is just a passing whimsy.

Yet for many scientists, the risks seem more seductive than repellent. "I'm not worried," said Dr. Charles Kimmel, the director of the Eugene lab. "If other people like fish, that's fine. If not, that's fine, too. In the meantime, I'm having a terrific time, and I think I'm making some contributions."

As a measure of the vitality of the fish field, young biologists are clamoring to study in zebra-fish labs. They are drawn to the versatility of the fish and its gorgeousness when viewed in molecular detail; they also see it as a magnificent career opportunity.

Half of Dr. Nusslein-Volhard's lab is still dedicated to fruit-fly genetics, but none of the young scientists applying to study with her wants to join that half.

"The post-doctoral applicants now go for fish, " she said. "They're afraid they're not going to get positions if they study flies because that field is perceived as being saturated."

The fish trade may only now be taking hold, but its history dates to the early 1970's, when Dr. George Streisinger, a geneticist at the University of Oregon, decided it was time to move on from his viral studies to an understanding of vertebrate development. He wanted to find an organism that he could manipulate with the powerful tool of genetics—the ability to alter genes and see their impact on the animal. The candidate had to be small, a quick breeder and able to thrive in a lab. As an avid fish hobbyist, he decided upon the zebra fish, purchasing his first experimental animal from a pet store.

In fashioning his new fish lab, Dr. Streisinger came up with several technical breakthroughs. Biggest among them was the ability to do a sort of virgin-birth trick with the fish. In the technique, semen is gently squeezed out of a male fish and irradiated with ultraviolet light, which destroys the genetic material in the sperm cells but leaves their fertilizing enzymes intact.

When the irradiated sperm is mixed together with a female's egg in a lab dish, the sperm has no genes to bestow but still manages to spark the growth of an embryo, spawning an infant fish that has inherited its genes solely from the mother, rather than from both mother and father.

That unnatural baby fish, called a haploid, is particularly easy to study because its genetic complement is known to be purely maternal. The offspring are feeble and cannot survive more than a few days, but before their demise a researcher can glean insights into their development.

Dr. Streisinger also devised a system for creating genetic mutations in fish, a critical tool for understanding genes and the proteins they produce. Mother fish are exposed to gamma rays that lightly scramble the DNA of the eggs they carry. As a result, the mother gives birth to mutants, which lack key genes necessary for health.

Studying the disruptive effect of a mutation in a vital gene is the best way to understand the normal task of the gene. Geneticists spend a lot of time looking for what they describe as "interesting" mutations—obvious and intriguing defects in an animal's anatomy or behavior that can be studied for clues to the healthy body or brain.

In 1984, just as he was beginning to gain wide attention for his fish work, Dr. Streisinger died of a heart attack at the age of 56. "When I heard he had died, I got very upset," said Dr. Nusslein-Volhard, who had been following his research. "I thought everything he had done may have died with him."

But Dr. Kimmel stepped in to pick up Dr. Streisinger's work. The fish facilities were expanded and fish care was honed to the greatest possible efficiency. The lab now houses about 10,000 zebra fish, grouped in separate subsections of tanks depending on, for example, the particular mutation they may have.

Feeding huge numbers of fish has proved no easy chore. "A lot of man-hours are spent keeping the fish happy and in breeding condition," said Charlene Walker, a senior research assistant who has been working with the fish since 1971. The fish are fed twice a day on an extravagant diet of live brine shrimp from San Francisco ("the very best we can possibly find," said Ms. Walker); a cheese-like bar of minerals; and commercial trout chow, pulverized for easy consumption.

At her lab, Dr. Nusslein-Volhard has discovered the ideal food for zebra fish is none other than fruit-fly larvae, with which she is amply supplied.

Researchers also work hard to get as many usable eggs as possible for their genetic machinations. When cued by the proper combination of light and darkness, females can be spurred to lay eggs every day, but they will eat many of their own eggs if given the chance. To prevent the loss, marbles are placed at the bottom of a breeding tank. The eggs drift down between them, beyond the reach of a fish's lips, and they are fetched later for fertilization in a lab dish.

For other experiments, males and females can be placed together in a tank and allowed to breed as they will, which they do in great thrashing swarms of females exuding eggs and males responding frenetically by expelling sperm. "I love to watch fish mating," said Dr. Hopkins. "It's one of my favorite activities. I feel like a pervert, but I have to say it's a joy."

Through their painstaking efforts, scientists are now able to breed fish with provocative mutations that cast light on fundamental questions of development. Some scientists are studying a mutant with the evocative name of spadetail.

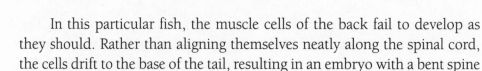

In this particular fish, the muscle cells of the back fail to develop as they should. Rather than aligning themselves neatly along the spinal cord, the cells drift to the base of the tail, resulting in an embryo with a bent spine and a spade-shaped tail.

Scientists find the mutant useful in their search for the genes that tell muscle cells what they are, where they belong and which motor neurons they are meant to link up with. Using the spadetail mutant, Dr. Judith S. Eisen of the Eugene lab has done an extraordinary series of microtransplantation experiments, shifting individual motor neurons from one part of the embryo to another.

The cells are stained a brilliant fluorescent red, allowing their progress and fate to be precisely tracked. She has learned that the neurons are pre-programmed to hook up with their target muscle cells, whether the right muscle cells are in the neighborhood or not.

"Even if you put the neurons in a new location," she said, "they do the right thing," sending out finger-like projections in a fruitless reach toward their muscle-cell partners.

The beauty of the zebra fish, she said, is that such single-cell transplants can be done. In similar studies of embryonic chicks or mice, "the best you can do is move around whole groups of cells," she said.

Another mutant zebra fish under intensive study is named Cyclops, after Homer's one-eyed cannibal. Because of its genetic flaw, that fish lacks a group of cells vital for guiding nerve cells in the developing brain. As a result, the brain fails to grow in the embryo and the eyes fall together into a single ocular mass.

With the aid of the Cyclops mutant, Dr. John Y. Kuwada, of the University of Michigan in Ann Arbor, is closing in on the series of signals that allow developing neurons to distribute themselves across the brain and to make their synaptic connections with one another.

Dr. Kuwada and others are also trying to isolate the genetic sequences that are disrupted in the mutant fish. That problem is proving particularly daunting. Thus far, no genetic map for the fish has been developed, so nobody knows on which chromosomes any of the fish's tens of thousands of genes may be located.

Zebra-fish experts know only too well that their specialty is, as Dr. Kuwada said, "very, very embryonic." The scientists talk wistfully of the day when they will be able to use the fish to comprehend the genetics of social behavior, like understanding the genes that cause fish to swim in schools, for example. But then they remember that they are spending hours of every day just scrubbing down the fish tanks, and they remember how unglamorous the life of a pioneer can be.

—Natalie Angier, November 1991

Catfish Slime Has Healing Agents

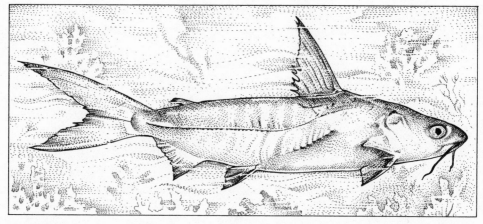

Glenn Wolff

Arabian saltwater catfish secretes a gel-like slime on its outer surface that helps heal wounds.

CATFISH SLIME, a gel-like substance secreted by the fish, has remarkable properties that help heal wounds, a team of American and Kuwaiti scientists has discovered.

The scientists made their discovery while studying marine life in the Arabian Gulf. Richard S. Criddle, a professor of biochemistry and biophysics at the University of California at Davis, said that when a local Gulf species of catfish is caught, it secretes a slime over its entire body.

"I have used it myself on cuts," he said. "They heal entirely in 3 days, instead of the usual 10."

Dr. Criddle said that his colleague, Jassim al-Hassan, a professor of biochemistry at Kuwait University, saw Arab fishermen rub the slime on cuts and scratches several years ago. It was around as folk medicine for a long time, Dr. Criddle said, "and has only been rediscovered by us."

American catfish, including freshwater species, and many other fish also secrete a similarly beneficial slime, but they tend to secrete it beneath their outer skin, Dr. Criddle said. The Arabian saltwater catfish, *Arius bilineatis,* which grows up to three feet long, is different in that it secretes the slime on its outer surface, making it more accessible and easier to isolate, he said.

A detailed analysis of the slime has turned up about 60 different proteins that are fundamental agents of wound healing in humans and other animals. One activates prostaglandins, substances that help initiate inflammation and pain responses to wounds. Others block bacterial growth.

The slime also contains a high concentration of the molecules that coagulate blood, forming clots that stop bleeding, and enzymes that accelerate cell division and the formation of new tissue.

The various substances in the slime trigger all sorts of chemical reactions under the skin, Dr. Criddle said, "bringing in white blood cells to break up broken tissue and clean it out, bringing in cells that start making repair products and then shutting the whole process down when the wound is healed."

The catfish probably developed the ability to heal rapidly, Dr. Criddle said, because a wounded fish bleeding in the ocean is likely to attract predators. The slime would also keep sores from contact with dirty water.

Catfish slime may eventually be used to accelerate wound healing in patients who do not heal well, such as diabetics and older people, and for burn victims, Dr. Criddle said.

Drug companies have made inquiries about the slime, but as part of their overall approach to developing better wound-healing products, he said. The complexity of duplicating the proteins in the proper balance might preclude synthesizing it.

—SANDRA BLAKESLEE, January 1988

3

FRESHWATER
FISH
IN PERIL

All over the world, freshwater fish face threats to their existence. Overfishing has reduced the famed sturgeon of the Caspian to pitiful numbers. Even in the United States, where several decades of environmental laws have helped clean up the water and protect endangered species, many species of fish are in grave danger.

As the homes of aquatic creatures, rivers are severely affected by what happens on their banks. Farming and logging are both bad news for fish. So are dams. Many American rivers are now no longer hospitable places for fish to thrive.

East Coast rivers lost most of their salmon long ago, and the West Coast has been following the same pattern. Another game fish, the trout, will be in serious peril if the dreaded whirling-disease parasite, which has already gained a foothold in the United States, cannot be contained.

Even where native species of fish have maintained a finhold, they are often threatened by invaders introduced by stocking programs or when an aquarium is flushed into a stream. When these invaders flourish, it is often at the local inhabitants' expense.

Under this battery of assaults, no less than a third of America's native fish are now considered rare, approaching extinction, or to have already vanished.

Bizarre Parasite Invades Trout Streams, Devastating Young Rainbows

RAINBOW TROUT and Montana's Madison River were made for each other. Bold and sometimes reckless, stouthearted and acrobatic when hooked, the pink-striped rainbow is the pre-eminent native American trout, and it has long thrived spectacularly in the powerful currents and chaotic, boulder-strewn riffles of what some people consider the finest wild trout stream in the contiguous 48 states.

"This was the classic rainbow water," Fred Nelson, a state fisheries biologist, said as he and two fishing companions paused to take in the scene: the sun glistening off the rushing river, the cloudless cobalt sky, the sharply etched mountain peaks in the distance. It almost made up for the fly fishers' total lack of success; no one had caught a trout. In fishing, that sometimes happens. But things are sadly different on the Madison these days. "It's real noticeable that you just don't catch rainbows like you used to," Mr. Nelson said.

In fact, the Madison's renowned population of magnificent wild rainbows is crashing, victimized by an exotic parasite that nobody yet knows just how to combat. In prime rainbow habitat, the species' numbers have plummeted recently to 10 percent of what they were. "Out here we say the vandals have broken into the cathedral of fly fishing," said Dr. Karl M. Johnson, a virologist who in the 1970's discovered the Ebola virus as a human killer and who now is one of many experts grappling with whirling disease. He moved here for the fishing, and his expertise has made him a central figure in the struggle to understand the scourge.

Whirling disease gets its name because the parasite, a tiny animal organism with a bizarre life history straight out of *Alien*, often drives young trout to a frenzy of frantic tail-chasing before they die. Not only the Madi-

61

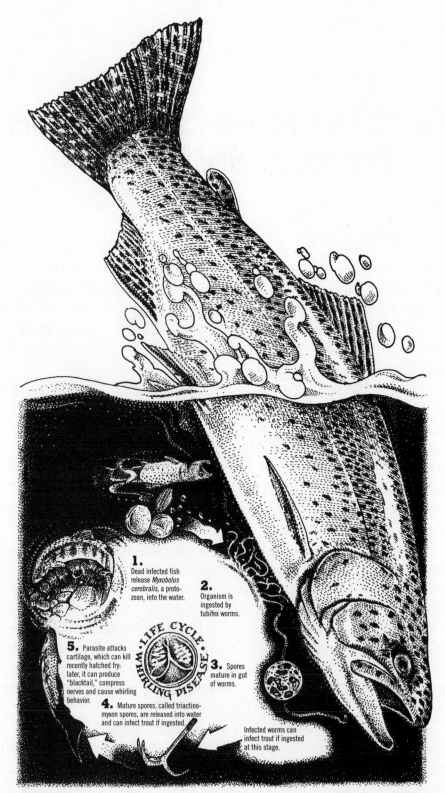

1. Dead infected fish release *Myxobolus cerebralis*, a protozoan, into the water.

2. Organism is ingested by tubifex worms.

3. Spores mature in gut of worms.

4. Mature spores, called triactinomyxon spores, are released into water and can infect trout if ingested.

5. Parasite attacks cartilage, which can kill recently hatched fry; later, it can produce "blacktail," compress nerves and cause whirling behavior.

• LIFE CYCLE •
WHIRLING DISEASE

Infected worms can infect trout if ingested at this stage.

Glenn Wolff

son is affected; the parasite has infected many streams in the Rocky Mountains, on both sides of the Continental Divide, and there are fears that it could eventually wipe out not only most strains of rainbows but also the cutthroat—the trout of Lewis and Clark—in much of the West. The cutthroat is an important forage item of many other animals, including the threatened grizzly bear.

The parasite has been documented in at least 20 states from coast to coast, including New York, although only a few have so far experienced trout population declines. Here in the Rockies, the infestation typically advances downstream 15 to 20 miles a year, and once infected, a river apparently is infected for good. No one knows for sure how far the infestation will spread. "I think we can only speculate," said Dr. Johnson, "but you probably have to start out with the reasonable assumption that sooner or later it's going to spread downstream" through the entire Missouri River watershed of which the Madison is a part. In many areas, he fears, rainbows "are going to get slaughtered."

Wildlife biologists say that what is happening here is a striking illustration of the way in which humans are stirring the evolutionary pot around the world. Rainbow trout never encountered whirling disease before the parasite was inadvertently transplanted here from Europe, and unlike their European cousins, the brown trout, they apparently have evolved no widespread immunity to it. Rainbow populations will either adapt, as the European cousins presumably did in ancient times, or die out. The West could be witnessing an especially dramatic instance of natural selection in action, said Richard Vincent, the regional fish manager for the state of Montana, whose studies of Madison River trout populations have made it one of the best-documented streams anywhere.

In nature, if enough members of a species carry genes that enable them to resist disease or any other environmental threat to survival, descendants of those members will eventually multiply and take over the population, which may then re-expand. If there are no resistance genes, the population may disappear. It is not yet known whether any rainbow strains are resistant to whirling disease, but some portents are grim: Biologists note that in some reaches of the Colorado River, the disease appears to have wiped out all but 1 percent of the population of young rainbows, bringing it to the verge of local extinction.

How it all comes out has important economic implications for the new economy of the West, based as it is not so much on mining, ranching and logging these days as on the attraction of its natural splendors for visitors, retirees and businesses.

One saving grace is that the Madison and other affected streams still have healthy populations of brown trout, equally magnificent game fish. In this evolutionary context it is dominant. Browns not only have adapted successfully to whirling disease, they also are acting as permanent, healthy host reservoirs from which the parasite can spread to the much larger populations of rainbows. The browns, European natives, are thus shouldering aside the American rainbows as the primary game fish of the Madison, at least for now. Browns are larger on average, too. The only problem is that there are only about half as many browns as there used to be rainbows, and they are wilier and harder to catch. Will anglers take their money elsewhere? Some fishing guides report a drop in business this year, but it is not clear whether whirling disease is responsible.

Most trout waters are still unaffected even in Montana, where, according to a preliminary report by a special state task force, the disease has been confirmed in 10 locations. The fear is for the long term, and the fate of the Madison has focused attention as perhaps nothing else might.

Rainbows are native to the Pacific watersheds of North America and Asia and were introduced into the Madison about a century ago. Browns from Europe were introduced at about the same time; Buffalo Bill Cody, the Wild West showman, is said to have planted one batch of browns. Two decades ago, Montana eliminated stocking on the Madison and other streams and began managing these waterways as wild trout fisheries. The result: Large populations of big, wild rainbows and browns blossomed and turned the state into a fly-fishing mecca.

Mr. Vincent was intimately involved in establishing the wild fishery, but in 1991 population surveys found that rainbow numbers in the upper Madison, where the rushing-water habitat is most favorable for the species, had dropped. From then until last year the population in the survey area shrank from about 3,300 to about 300, and the decline is moving steadily downstream. Browns, meanwhile, have remained stable at about 1,500 per mile. "It was a shock to all of us" when the latest numbers came out last fall, says Tom Anacker, a lawyer in Bozeman, Montana, who is a founder of a newly organized Whirling Disease Foundation and member of the state task force.

It was not until December 1993, however, that Dr. Beth MacConnell, a histologist and pathologist for the United States Fish and Wildlife Service in Bozeman, identified the cause as whirling disease. The confirmation of the parasite's presence in Montana and Colorado raised alarms as they had not been raised before. Until then, "it wasn't thought you'd lose a lot of fish" to the disease, said Peter Rafle, a spokesman for Trout Unlimited, an angling and conservation group that has taken a leading role in spreading the alarm.

The disease is produced by a microscopic water-borne protozoan, harmless to humans, called *Myxobolus cerebralis*. Its spores are released into the water when infected fish die and decompose or are consumed and excreted by predators and scavengers. The spores may live in mud on stream and lake bottoms for up to 30 years, until they are ingested by a tiny, red, thread-like host called the tubifex worm. In the worm's gut, the spores are converted to the form of the parasite that affects the trout.

This form's true name is Triactinomyxon, "but we call it the grappling hook," said Dr. Johnson, who describes it as having "a head loaded like a shotgun shell" containing 64 infectious spores each about the size of a red blood cell. Appendages that look like grappling hooks make contact with the gills or skin of a fish, and the spores are injected. By eating a single tubifex worm, a trout can be infected by thousands of spores.

The spores attack the developing cartilage of young fish; older fish, in which cartilage has turned to bone, appear unaffected. How the spores kill young fish is not entirely clear. Dr. Johnson believes that the fish "just puts tremendous energy into attempting to kill off all the spores" and can no longer deal with all the other stresses it must cope with in the wild environment. The tail-chasing, he said, apparently results from an inflammatory response when the parasite reaches cartilage in the head, where the fish's balance organs are located.

The parasite vaulted across the Atlantic, scientists believe, in a shipment of frozen trout fillets from Denmark in 1956. Scientists are not sure how it got into wild American waters, but the disposal of contaminated fish parts is suspected.

Only recently has the disease become a serious problem in America. It has cropped up most prominently in trout hatcheries and apparently spread into the wild when hatchery fish were released. Dr. Johnson, who heads the scientific subcommittee of the Montana state task force on whirling disease, believes that the improved technology of transporting fish may have accelerated the trend.

In Montana, which relies primarily on the reproduction of wild fish, no parasites have yet been discovered in hatcheries. No one knows just how they got into the Madison and other waters, but illegal stocking is suspected.

In Colorado, where most drainages have been stocked from hatcheries, the impact has been widespread. Except for one or two drainages in western Colorado, all drainages in the state have been affected, said R. Barry Nehring, a state fisheries biologist who is the co-author of a definitive report on whirling disease in that state. The spread of the parasite has so far caused population declines of rainbow trout in only six rivers or streams, but one of them is the Colorado.

The parasite has been found in some New York state hatcheries and a few widely scattered streams, including a tributary of the fabled Beaverkill-Willowemoc system, the cradle of American fly fishing. All fish in the infected hatcheries were destroyed, and monitoring of streams is continuing. So far no population crashes have taken place, but it is unclear whether New York will escape or whether the full force of the outbreak is yet to come. "We're not sure where we are, frankly," said Phil Hulbert, a biologist who heads the Department of Environmental Conservation's cold-water fisheries unit.

While many questions remain unanswered, it seems clear that the whirling disease parasite is here to stay. "You're not going to get rid of the brown trout and the worms, which means you're going to have it forever," said Dr. Johnson. The big question, he said, is whether "we are going to be able to do anything rather than document it and watch the population die."

Montana is developing a list of practical suggestions for trying to contain the parasite. For example: Remove all mud and aquatic plants from boats, trailers and waders and boots. Do not transport live fish, an illegal activity in this state in any case. Do not dispose of fish or fish parts where they can reach any water supply.

One possible long-term strategy is to introduce in the Rockies a strain of California rainbows known to be resistant to a native parasite similar to *M. cerebralis,* on the theory that there is a reasonable chance those trout will be resistant to whirling disease as well. Dr. Johnson says this possibility offers at least "a shot" at dealing with the disease.

Mr. Vincent favors letting nature take its course in the hope that the few remaining Madison rainbows will be resistant and that the population can be rebuilt from them. Dr. Johnson has his doubts. It may be, he says, that the survivors are not resistant but were merely lucky in not contracting the disease; and even if they are resistant, he says, it would take a long time for the population to rebuild naturally—"beyond our lifetime and our children's."

—WILLIAM K. STEVENS, September 1995

Pollution Threatens
World's Caviar Source

AFTER TWO MILLION YEARS of a most celebrated existence in the Caspian Sea, the sturgeon—and caviar, its revered offspring—have been pushed by pollution, corruption and greed to the edge of extinction.

More than 90 percent of sturgeon and 95 percent of all black caviar come from the Caspian, which despite its name is the world's largest lake. A full-grown beluga, the largest member of the sturgeon family, can weigh a ton and carry more than two million eggs, although its relatives, the sevruga and osetra, are more common and not nearly as big.

Sturgeon were once so plentiful here and on the southern coast of Russia that fishermen still remember a time when it was difficult for boats to sail between them. The catch has fallen to less than 1 percent of what it was 50 years ago. In fact, some of the rarest species of sturgeon are already considered extinct.

"After millions of years of quiet and beauty, this sea and everything in it has rapidly started to die," said Arif E. Mansurov, the Azeri Minister of Ecology. "Three generations of evil, ignorant people have been able to wipe away millions of years of nature's work. Unless something changes soon, we will be the last generation to taste this caviar. I cannot say honestly that we alone are in a position to do a thing about it."

The Caspian suffers from many maladies, and they have come together in a particularly destructive way to imperil the fish that inhabit it. Baku, Azerbaijan's capital, about 100 miles north of here, was the world's greatest oil town at the turn of the century. Thousands of derricks cover this once majestic coastline and a thin, permanent film of oil, about one-quarter inch thick, has formed on the top of the water. In one of history's most brazen displays of disregard for nature, Stalin dammed the Volga, draining millions of gallons for reservoirs and huge electrical plants. This has altered the flow of water into the Caspian and added tens of thousands of tons of heavy metals, chemicals, raw waste and other pollutants to it each year.

It has also made it impossible for sturgeon to swim hundreds of miles up the Volga and Kura rivers, as they did for centuries, to spawn. In addition, for more than a decade the sea has been rising, and various theories suggest that this may have to do with the oil itself, which may prevent evaporation, or with global warming or with some kind of natural global cycle. As a result, huge silt barriers have grown up at the mouth of the rivers, particularly the Kura, which enters the Caspian here.

That means that when the fish do leave the sea to spawn they rarely make it back alive. Scientists predict the sea will rise at least seven more feet by the end of the decade, which would be beneficial for the fish but would also wipe out dams and hundreds of thousands of acres of farmland in three countries, and threaten communication towers, railroads and thousands of homes. The few fish farms created here and in Russia in the last 25 years to help replenish the diminishing stock have already been destroyed by the high waters.

The environmental disasters that have befallen this region pale, however, when compared with the political realities that have accelerated the slaughter of sturgeon.

Ten years ago, two nations bordered the Caspian: Iran and the Soviet Union. The sea, though polluted and mismanaged, was not overfished, because the Soviet authorities punished poachers severely. Today there are five countries on the rim of the sea: Russia, Azerbaijan, Turkmenistan, Iran and Kazakhstan. Iran is by far the most stable and the least threatening to the fish population.

The others, agree scientists, politicians and fishermen, are governed by chaos—when they are governed at all. Poachers now take as much as 10 to 20 percent of the annual catch from Russia and Azerbaijan, as a quick trip to any black market in this nation, or in southern Russia near the caviar capital of Astrakhan, will demonstrate. Medium-quality caviar that costs six dollars an ounce in Moscow and five times that in New York sells for five dollars a pound in the markets of Baku.

Without the former Soviet Union to enforce its brutal prohibition against poachers, the bounty of the sea is treated like a ready source of hard-currency exchange by crooks, peddlers and desperately poor governments like that of Azerbaijan.

"It is hard to imagine the environmental and economic damage from poaching in the Caspian Sea," said Alexander Frolov, a Russian Interior Ministry expert on organized crime who runs the ministry's department of special operations. He has established a new squad, complete with assault weapons, rapid-reaction units and night-vision goggles, to arrest the poachers. "This has become a huge criminal industry," he said after the highly publicized closing of an illegal caviar factory in Astrakhan. "There is a lot of money to be made."

As a general rule, 10 percent of the weight of a female sturgeon consists of eggs (although pollution-related diseases have helped lower the figure in the Caspian). If a 700-pound fish delivers 70 pounds of exquisite caviar—the type that can sell for as much as $100 an ounce in restaurants in New York or Paris—then that fish would be worth more than $100,000 to high-price retailers (using the kind of admittedly imprecise math on which illegal drug sales are usually based). Each female fish brings thousands of dollars on the open market. In Neftechala, where the average monthly income for a worker in the single state factory is $10, that kind of money is impossible to ignore.

The giant beluga takes nearly 20 years to reach maturity, and can live a century or longer. For these reasons, of all the sturgeon it has been the most severely affected by the environmental and economic anarchy that has taken over the Caspian. Poaching has become so widespread that the wholesale price of caviar on the world market has begun to fall. And beluga, the most prized catch of all, is disappearing faster than the rest.

"It's crazy what is happening here, just heartbreaking," said Abdul G. Kasimov, head of the department of hydrobiology at the Caspian Sea Research Center in Baku. "We need a complete ban on fishing for sturgeon in the sea. We agree to stop it and so does Russia. But our government is not capable of doing a thing against pirates. And we can't undo the pollution that has made so many of the fish sick. We are just too poor to do what we need to do."

"The problem with sturgeon is critical, I agree," said Paulo Lembo, the chief representative of the United Nations Development Program in Baku. "But when we have a million refugees in the country and a water supply system that is not even capable of delivering basic drinking water, it is difficult to raise money to save caviar."

There are rules to protect the vital cash crop, but they are rarely observed. In order to keep a level population, fishermen should take fewer than 50 percent of the fish as they leave the sea to spawn; these days, the take is close to 90 percent. And fish that are not harvested often die of natural causes or in their attempts to return to the sea.

In the end, it is the pain inflicted on the environment that has hurt the fish the most and caused the gravest concern for the future, Mr. Kasimov said. Increasingly, fishermen here say they have to throw sturgeon back because they are clearly infected with a serious virus or suffer from myopathy disease, which causes a process of degradation of the muscular and sexual tissues, preventing reproduction.

The rising waters, which have already wiped out 6 percent of the territory of western Kazakhstan and nearly as much of coastal Azerbaijan, would normally be a good thing for the fish. More water would bring more types of food and a richer diversity to the sea. And in the past, the sea has fluctuated greatly. It fell consistently from the late 1920's into the 1970's. By 1977 it had dropped almost 100 feet below sea level, its lowest point in hundreds of years. After World War II, however, both Iran and the Soviet Union began to carpet the newly dried sea bed with factories, villages and power lines.

Now the sea is taking it all back again. The extreme level of chemical and oil pollution in the water—and the threat of more if the waters destroy oil platforms—have made the rise deadly to life within the sea.

"Nature has certain laws and the farmers and fishermen here always respected them," said Urkhan K. Alekperov, vice president of the Azerbaijan Academy of Sciences and an expert on the life of the sea. "For thousands of years our villagers knew not to build on the sea's edge, not to live there. They knew that one day the sea would come back. But in this century nobody listened to the past. We are just starting to suffer the consequences."

—MICHAEL SPECTER, June 1994

River Life Through U.S. Broadly Degraded

TWO DECADES OF FEDERAL CONTROLS have sharply reduced the vast outflows of sewage and industrial chemicals into America's rivers and streams, yet the life they contain may be in deeper trouble than ever.

The main threat now comes not from pollution but from humans' physical and ecological transformation of rivers and the land through which they flow. The result, scientists say, is that the nation's running waters are getting biologically poorer all the time and that entire riverine ecosystems have become highly imperiled.

Dams disrupt temperature and nutrient patterns on which organisms depend. Countless river and stream channels have been straightened, eliminating the meandering course on which rivers depend for their ecological variety. Repeated diversions of water from a river's flood plain can decimate populations of fish that spawn there. Sediments from farming run into streams and suffocate many small forms of aquatic life. Vacationers who cut down trees to improve the view in front of summer homes may erode stream banks. The stream then carries more sediment and becomes wider, shallower and warmer, making the water unfit for many vital organisms.

"If you take a drive out into pretty, rolling farm country, nobody thinks of the farming activity as habitat destruction," says Dr. J. David Allan, a freshwater ecologist at the University of Michigan. "But the transformation of the landscape by agriculture is taking its toll" on life in rivers and streams, as are urban and suburban development and the spread of exotic, disruptive species of aquatic life.

The transformation, says Dr. Allan, is far more destructive to aquatic life than are spills of oil or toxic chemicals. For all the onetime harm they may cause, these spills have relatively little long-term impact. And because

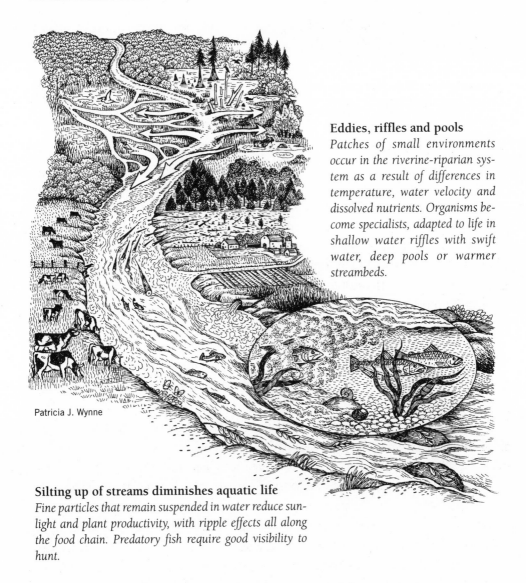

Eddies, riffles and pools
Patches of small environments occur in the riverine-riparian system as a result of differences in temperature, water velocity and dissolved nutrients. Organisms become specialists, adapted to life in shallow water riffles with swift water, deep pools or warmer streambeds.

Patricia J. Wynne

Silting up of streams diminishes aquatic life
Fine particles that remain suspended in water reduce sunlight and plant productivity, with ripple effects all along the food chain. Predatory fish require good visibility to hunt.

the transformation is so much a part of deeply entrenched patterns of land and water use, it is also far harder to deal with.

Dr. Allan lays out the threat to riverine organisms and ecosystems in an article in a recent issue of the journal *BioScience*.

A 1990 study by Larry Master of the Nature Conservancy found that in North America, 28 percent of amphibian species and subspecies, 34 percent of fishes, 65 percent of crayfish and 73 percent of mussels were imperiled in degrees ranging from rare to extinct. The comparable figures were 13 percent for terrestrial mammals, 11 percent for birds and 14 percent for land reptiles.

In the West, where dams and the introduction of exotic species are common, the situation is particularly acute. Of 30 species of native fish in Arizona, 25 are listed as threatened or endangered, according to Dr. W. L. Minckley, a zoologist at Arizona State University.

The biotic impoverishment goes beyond the loss of individual species, however. Many rivers, Dr. Allan wrote, contain few or no endangered species, yet there are so few representatives of each species present that the ecosystem's functioning is impaired. Scientists do not know at what precise point this thinning of life causes an ecosystem to disintegrate. But "it's like an airplane wing," said Dr. John Cairns Jr., an environmental biologist at Virginia Polytechnic Institute, explaining, "if you keep pulling rivets out, the wing is going to go."

Among other benefits, riverine ecosystems create breeding grounds for commercial fisheries, carry nutrients to them and support multimillion-dollar recreational activities. In concert with wetlands, they regulate the flow of water, releasing it more slowly in flood times so that more will be left for dry times.

Few if any major river systems are unaffected by the threat to ecological integrity.

Sediment from farm fields, for instance, has clouded the mighty Mississippi, making it more hostile to many organisms. Levees prevent the sediments from settling out naturally on the Mississippi Delta. Instead, they are channeled directly to the continental shelf. This contributes to a sinking of the land in southern Louisiana and releases so many river-borne nutrients into the Gulf of Mexico that plankton growth is stimulated. The plankton use up oxygen when they decay and die, and scientists fear this oxygen depletion may harm Gulf fisheries.

The Colorado River south of Lake Mojave has been so altered by disruption of water flow and the introduction of exotic fish species, Dr. Minckley said, that it has become the first major river in North America with no native fish left.

Dams on the Columbia River have so interfered with salmon migrations that one variety of Columbia salmon has been listed by the government as endangered. Another has been declared threatened, and five more have been proposed for listing.

All three of these watercourses appear on a 1992 list of North America's 10 most endangered rivers compiled by American Rivers, a Washington-based conservation organization. Others include the Alsek and Tatshenshini river system in Alaska and Canada, the Great Whale River in Quebec, the Everglades, the American River in California and the Penobscot in Maine.

The list is rounded out by the Beaverkill and the Willowemoc, legendary Catskill trout streams where American fly-fishing was born, and Montana's Blackfoot, the putative setting of the movie *A River Runs Through It*.

Habitat in lower stretches of the Beaverkill-Willowemoc system is threatened by developers' cutting of streamside vegetation. The Blackfoot has become so degraded by timber cutting, agriculture, water diversions and mining activities that the moviemakers were forced to move to another location.

Kevin Coyle, the president of American Rivers, describes "the four horsemen of river destruction" as dams, diversion of water, alteration of channels and land development.

Dams trap nutrients and keep them from flowing downstream. Perhaps more devastating, they alter the temperature of downstream water, making it either too cold or too warm and thus annihilating whole populations of insects vital to the riverine food web. One dam might not be so bad, but many dams on the same river, as is common in the West, repeatedly interrupt the river's natural functioning.

Diversion of water for human use, also widespread in the West, has simply dried up many rivers and streams for much of the year, with the result that their ecosystems are, in Mr. Coyle's words, "ghosts of what they used to be."

The straightening, diking and redirection of river channels, common across the country to control floods and convert flood plains to cropland, housing and highways, reduce the variety of habitats critical to biological diversity.

Land development often denudes stream and riverbanks of vegetation, eliminating the vital transition between the river and the uplands. Draining land for farming or development causes water to flow more rapidly into the river channel than it naturally would. This leaves less water to percolate into the river in drier times.

If the river channel has been straightened as well, water draining from the land moves more efficiently, producing more powerful floods. These carry the increased sediments from farming and development farther, choking organisms and ecosystems well downstream.

On top of all this, legions of exotic species have been introduced into running waters. Some, like the zebra mussel slowly spreading across the country, have appeared by accident. Others, like fish imported to provide sport or to clean vegetation from the waters, have been introduced on purpose. Together, Dr. Allan said, they have significantly reduced biological diversity through predation, alteration of habitat, introduction of diseases or parasites and interbreeding with native organisms.

Such ecological tinkering can unexpectedly cascade through the water, onto the land and into the economy as well. In one instance, fishery managers in Montana introduced opossum shrimp into Flathead Lake and its associated river systems, hoping the shrimp would provide forage for kokanee salmon that were the basis of a thriving tourist industry.

Instead, the shrimp consumed zooplankton that were the staples of the kokanee diet. The kokanee population collapsed. Bald eagles and grizzly bears that once congregated at the rivers to feed on salmon disappeared, as did tourists who had come to see them.

Once invasive species have established themselves, said Dr. Allan, it may be impossible to eliminate them. The other main causes of biological impoverishment seem only a little less intractable. Even so, Dr. Allan, Mr. Coyle and others say much can be done.

American Rivers advocates a three-pronged strategy: saving the headwaters of the major rivers, which for the most part are already publicly owned; protecting and restoring riparian zones by replanting green strips along rivers; and working with governments to regulate water discharges from dams so they disrupt ecosystems less. Federally controlled dams are also being examined for their environmental effects as their hydroelectric licenses come up for renewal.

A number of scattered efforts to restore rivers and streams are being undertaken. Restorationists have become expert at restoring streams for game fish like trout, Dr. Allan noted. What is needed now, he said, is a comparable effort to restore habitat for the full panoply of riverine organisms.

An ambitious effort along these lines involves the Kissimmee River in Florida. To control flooding, the Army Corps of Engineers basically turned the twisting, 103-mile-long river into a straight canal, largely destroying the riverine-riparian ecosystem. Now, after a successful demonstration project, the state of Florida and the corps hope to restore the river's twists and turns—and its ecosystem.

Broader restoration of this sort is still in its infancy. But as fragile as riverine ecosystems are, Dr. Allan points out, they are also remarkably resilient. They tend to repair themselves once the causes of their impoverishment are removed.

So, he says, all is not lost.

—WILLIAM K. STEVENS, January 1993

Storm Swirls Over Aboriginal Salmon in Maine's Rivers

NO LESS THAN THE EAGLE, the wolf and the bear, the bullet-sleek Atlantic salmon has inspired human wonder. It captivated the Romans, who named it "salio" (the leaper), and Izaak Walton, who in 1653 crowned it the king of freshwater fish. The species even figures in Cro-Magnon cave art.

In America now, just as the bald eagle, wolf and grizzly bear are imperiled in the contiguous 48 states, so the original wild salmon of New England has come dangerously close to extinction, and some geneticists say it may already have vanished for all practical purposes.

Whether the "aboriginal" salmon has disappeared or not is at the center of a political storm in Maine, where the federal government proposes to list what its biologists say is "the last known wild remnant of U.S. Atlantic salmon" as officially threatened under the Endangered Species Act. State officials, arguing that aboriginal salmon are no longer genetically identifiable, vehemently oppose the listing.

The wild salmon has generally declined throughout the North Atlantic basin in recent years, but its situation in the United States is desperate, fisheries biologists say. Despite longtime efforts to restore the fish to New England rivers through restocking from hatcheries, a mere 1,500 to 3,000 salmon—down from 6,000 to 8,000 a decade ago—are estimated to make the annual spawning run to those rivers from feeding grounds off Greenland. Historically, as many as 500,000 made the trip.

Of those that do make spawning runs today, say federal biologists, only about 500 are truly wild. That is, they still carry a significant part of their unique genetic inheritance, developed over millenniums of reproductive isolation. Since salmon return to the stream of their birth to spawn, each river has, or at least had in the past, its own genetically distinct fish. The remaining 500 fish, the targets of the federal proposal, spawn and mature in seven Maine rivers: the Sheepscot, Ducktrap, Narraguagus, Pleasant, Machias, East Machias and Dennys.

But a study conducted by scientists for the state concluded that the seemingly wild fish have become so genetically swamped by interaction with hatchery fish that their distinctiveness has been lost. If so, the proposal to list them may be too late.

"There are no individual fish that retain the genetic integrity" of pre-colonial times, said Dr. Irv Kornfield, a population geneticist at the University of Maine at Orono, who headed the state study. While some original genes may remain, he said, they do not add up to the significant legacy cited by the federal biologists. "Either we have a tiny bit left or there's none," he said. If there is none, the overall salmon gene pool has been correspondingly weakened.

This is a case "where reasonable men can disagree," said Paul Nickerson, endangered species coordinator for the United States Fish and Wildlife Service's Northeast region in Hadley, Massachusetts. "I don't think any of us says the original salmon that was here before European settlement still exists." But, he contended, enough of the original genetic material lives on in the Maine fish to warrant federal protection.

The state of Maine objects "in the strongest possible terms" to listing the salmon as threatened, Governor Angus S. King Jr. has told federal officials. The state argues that conservation measures already under way are beginning to bring the salmon back. Officials also fear that listing the salmon as threatened could bring federal regulations harmful to blueberry growers who draw water from the salmon rivers and to aquaculturists who artificially raise salmon for market in ocean pens just off the mouths of the seven rivers. The state is pressing the federal officials to accept a state-devised conservation plan in lieu of listing.

Some conservationists also question whether listing is necessary. They cite the actions already under way and worry that many local citizens, who have expressed strong anti-federal sentiments, would withdraw their support for conservation measures in the event of a listing. But others argue that without the spur to action provided by listing, progress in restoring the salmon will be too slow.

"The state has failed and it's time to try something new," says David N.

Carle, associate executive director of a Massachusetts-based conservation group called Restore: the North Woods. A petition by this group forced federal officials to consider the listing and catapulted the issue to a new level of urgency. The group has favored declaring all salmon in New England endangered, including the progeny of hatchery fish.

The creature at the center of the argument, *Salmo salar,* has long been a human icon. Hardy and muscular, it can jump 12-foot waterfalls that its cousins, trout and Pacific salmon, cannot. And while all Pacific salmon die after spawning once, the Atlantic salmon can make several spawning runs, typically two or three. Those that survive to reproduce many times can reach majestic proportions, weighing more than 50 pounds, although five to 20 pounds is more usual.

In North America, the salmon's spawning range historically began with the Housatonic River in Connecticut and extended east and north through the Canadian maritime provinces. But by the early 1800's, the fish were gone from rivers like the Connecticut, driven out by dams, fishing and pollution.

For years, biologists in New England have been trying to restore them to a few rivers like the Connecticut and Massachusetts' Merrimack.

Progress has been slow. Self-sustaining populations—those not dependent on the stocking of hatchery-raised fry, or baby fish—have yet to become established. It simply takes a long time, given that salmon take several years to reach spawning age and that relatively few live that long, said Dr. Steve Gephard, supervising fisheries biologist for the state of Connecticut.

This year, 20 years after restoration began, about 260 spawners returned to the Connecticut River. These are not enough to insure the establishment of a new, self-sustaining population. In an attempt to speed the process, biologists are stripping eggs from these fish, fertilizing them, and raising fry in hatcheries, where many, many more can be produced than in the wild.

Small populations like those in New England are especially vulnerable to overfishing and natural disasters at sea. Curbs have been placed on commercial fishing for salmon, and a decade ago the outlook seemed bright: Salmon returning to the seven Maine rivers numbered more than 2,000 in some years.

In the early 1990's, however, the species at large entered a general decline, even as commercial fishing was further restricted. No one knows whether the cause was climatic change, a decrease in the shrimp and fish that salmon eat, a natural fluctuation in salmon population, or something else. In any case, salmon began disappearing into what biologists are calling a "black hole" in the Atlantic.

Like the wolf, bald eagle and grizzly, the salmon is relatively abundant north of the contiguous 48 states. Salmon populations have nevertheless declined there. For instance, 200,000 to 220,000 fish returned this year to the Miramichi River in New Brunswick, the most prolific Atlantic salmon stream in North America. That compares with about half a million 30 years ago and a million historically, according to Bill Taylor, president of the Atlantic Salmon Federation, an international conservation group based in St. Andrews, New Brunswick.

While the population plunge has depressed Canadian and European salmon stocks, it has put wild American salmon in critical condition. Their plight was exacerbated, biologists say, by Maine's former practice of stocking the seven rivers with hatchery fish, which competed with and genetically diluted the wild ones.

"It probably did more harm than good," said Edward Baum, a fisheries biologist who heads the state's salmon recovery program.

Since 1992, the state has been trying to build up the wild population in each of the seven rivers by stocking it with fry raised in hatcheries from eggs produced by wild fish found only in that particular river. Mr. Baum said this year's returns of adults from the sea, the 500 fish, represent at least a 25 percent increase from last year. Since the program "is just starting to pay off," he said, he believes it would be best to hold off a bit on listing.

Federal officials, however, say the measures taken by the state are the reason they are proposing to list the salmon only as threatened, rather than giving it the more serious label of endangered. If the measures fail, say the officials, the fish would be declared endangered. This would mean much more stringent federal involvement.

Under a listing of threatened, the state could retain its leading role in managing the salmon population. In response to the listing proposal, the state has prepared a conservation plan intended to prevent escaped salmon from the offshore aquaculture pens from genetically swamping wild fish, and to protect salmon habitat from pollution, agricultural water withdrawal and tree cutting. Catch-and-release angling would be permitted. The plan was worked out by state officials, scientists, industry and some conservationists, including the Atlantic Salmon Federation.

The more serious "endangered" listing, should it ever come, would prohibit angling for salmon on the streams in question. Proponents of the proposed state conservation plan say this would drive away anglers who have been mainstays of the conservation effort.

Mr. Carle rejects the state's conservation plan as inadequate, not least because, he says, it would be inadequately financed. Federal funds for salmon recovery would accompany a threatened listing. But Governor King insists that the government accept the state plan in lieu of any listing; otherwise, he said in a letter to federal officials, state cooperation will be withdrawn and legal action will be taken. The federal officials could accept the plan and still list the salmon.

A decision on the listing question is expected in midspring. In the meantime, the government is looking for a way to protect the salmon without igniting a rebellion in Maine. Mr. Nickerson of the Fish and Wildlife Service said he was worried that a lot of energy now being put into conservation would instead "go into fighting the Feds" if the salmon is listed.

"And the salmon," he said, "doesn't win there."

—WILLIAM K. STEVENS, December 1996

Dwindling Salmon Spur West
to Save Rivers

HOOK-JAWED, WILD AND STILL MAGNIFICENT after its taxing and hazardous struggle up the Rogue River from the ocean 125 miles away, the big coho salmon was beginning to don his scarlet mating colors, just as his kind has done every November for thousands upon thousands of years. But far fewer fish than made the trip just a few years ago are being recorded these days by the video camera trained constantly on a narrow underwater passage through which all migrating salmon must pass to get around Gold Ray Dam.

Lately, annual runs of wild coho on the Rogue as they head for their natal waters on tributaries upstream have been about 80 percent below average. Federal officials are expected to announce soon whether they will declare the coho an endangered species throughout much of the Pacific Northwest. Fisheries biologists say, in fact, that most strains of Pacific salmon—as valued a totem in this region as the bald eagle is in the nation at large—are in deep trouble as a result of overfishing at sea, natural climatic factors and destruction of freshwater habitat where the big fish spawn and where their progeny grow large enough to return to the ocean.

The wild coho population "is in pretty poor shape," Jerry MacLeod, a state fisheries biologist who directs Oregon's new watershed health program in southwestern Oregon, said as he gazed at the salmon flashing by outside the glass of the viewing station at Gold Ray Dam.

Bright spots are beginning to appear in this gloomy picture, however. Salmon that successfully negotiate pollution, unnaturally warm water and obstructions like dams are about to discover newly improved spawning streams like Little Butte Creek in the foothills of the Cascades, which flows into the Rogue near here. More than anything else, the deteriorating condi-

83

tion of these streams has seriously limited the ability of wild salmon, mainly coho and chinook, to rebound from record low numbers. Now governments, environmental groups, local volunteers, out-of-work loggers and even lumber companies are pitching in to restore spawning and rearing habitat to the complex conditions that salmon require.

To reproduce, salmon must have clean beds of gravel in which to spawn, and the hatchlings must have deep pools of clean water with a lot of nooks where they can hide from predators while waiting to dart into the current to seize aquatic insects, their main food. Under natural conditions, fallen trees create a stream habitat offering "that nice combination of holding, hiding and feeding; it's really neat stuff," said Dr. Ken Cummins, the author of a recent study of salmon habitat conducted for the state of Oregon by the Center for the Study of the Environment, a private research organization in Santa Barbara, California.

Over the years, people have removed many of the fallen trees, sometimes in the mistaken belief that migrating salmon could not get over them. As a result, the random assemblies of alternating riffles and pools have been converted to shallower, simplified, more uniform channels with few if any fish. The deliberate straightening of some channels has added to the simplification. Sediment let loose by logging, agriculture and development has choked some spawning beds. The cutting of trees along the stream bank—the riparian zone—has deprived some aquatic insects of the leaf litter on which they feed. By eliminating shade, tree removal has also allowed the water to warm, reducing the supply of microorganisms that other aquatic insects eat. Dr. Cummins found that long-term salmon production plummets under these conditions and takes three or four decades to recover after trees reappear and are allowed to grow to maturity.

Happily, freshwater habitat is the element in the equation of salmon endangerment that people can do the most about. Early results are already apparent on the Little Butte, the key spawning and rearing stream of the Upper Rogue watershed.

The area still looks wild and pristine. One raw and rainy day last week, wisps of cloud nestled between the crests of steep hills covered with Douglas firs, while early snow dusted some hilltops. But long reaches of the stream have been ecologically undone by the cutting of streamside trees and removal of fallen logs. Last summer, in what promises to be one of the first of many such attempts to re-create nature in the watersheds of the North-west, Mr. MacLeod's team brought in huge tree trunks to replace those that had been removed from the stream and anchored them in natural positions.

In just a few months, the scouring movement of water over, under and around the logs has already converted some parts of the stream bed from shallow, open riffles to shoulder-deep pools. After the coho now on their way to the Little Butte spawn this month, the hatchlings will find an invit-ing nursery. A test of this strategy in another stream increased the number of surviving young salmon a thousandfold, said Mark Grenbemer, a mem-ber of the watershed restoration team.

Upstream a bit, where the Little Butte passes through federally owned land, the United States Forest Service has replaced big logs in the stream and planted new trees along the bank to provide a long-term supply of shade and pool-forming logs. The Oregon state restorationists will take this step next summer on some private lands as well. The riparian zone is also to be fenced off from cattle, which pollute streams, destroy bankside vegetation and cause stream banks to crumble. For much of the work, local people, including out-of-work loggers, are being hired at a total cost of about $70 million this year and next.

Little Butte is one of several priority watersheds designated under both state and federal restoration plans. Oregon is committing proceeds of its state lottery to the effort. Federal funds are being supplied to both state and fed-eral efforts under the Clinton Administration's Northwest forest plan, devised originally to solve the politically incendiary spotted owl dispute. But salmon have now become its main focus, not least because unlike spotted owls, they enlist broad support in these parts. The Administration is extending the watershed restoration program well outside the range of the spotted owl.

"I started out thinking it was a spotted owl problem, but found that at bottom it was a salmon problem," said Interior Secretary Bruce Babbitt. Harmful land-use practices and mismanagement of ecosystems, he says, are most strikingly reflected and greatly magnified by their impact on streams. And in this region, the health of the salmon population is the clearest indi-cation of stream and watershed health: Restoring salmon habitat necessar-ily restores the biologically rich riparian ecosystem as well.

Various surveys in the last four years have documented the drastic decline and in some cases disappearance of many populations of Pacific salmon. Wild salmon populations, because they spawn in separate watersheds from one another, are isolated and thus genetically distinct. A 1991 study by the American Fisheries Society in the three West Coast states and Idaho found that of 400 distinct stocks of salmon, steelhead trout and sea-run cutthroat trout, 214 were at risk of extinction and 106 were already extinct.

A 1993 study by the Wilderness Society found that the coho—historically, along with the chinook, the most widespread, abundant and economically important species—had disappeared from 55 percent of its historic freshwater range in the same four states and was either threatened or endangered in another 33 percent. The status of sockeye and spring-run chinook was almost as depressed.

This year, the National Marine Fisheries Service reported that West Coast ocean catches of coho fell to only 292,000 in 1993 from 5.3 million in 1976, a decline of 95 percent. Chinook, the agency reported, had declined by 80 percent from 1988 to 1992. Commercial offshore fishing for coho has been banned, and for chinook has been sharply restricted.

An abundance of salmon and sea-run trout in Alaska may assure that none of these species will go extinct globally any time soon. Alaskan abundance also assures that the supply of salmon for market is substantially unaffected. But biologists fear that coho, chinook and sockeye, along with steelhead, a sea-run species of rainbow trout, could be erased from all but a small fraction of their former range.

In addition to its imminent ruling on the status of the coho, the National Marine Fisheries Service will rule later on whether to consider the steelhead officially endangered in much of its range. In both the coho and steelhead cases, the habitat restoration efforts will be taken into consideration in deciding the status of the two species and what remedial measures, if any, are required, said Merritt Tuttle, a fisheries biologist who is the senior policy analyst in the agency's office in Portland, Oregon.

In both freshwater and salt, the fish are vulnerable to many threats. In recent years, salmon numbers are thought to have been reduced by an extended drought and a drop in ocean food supply brought about by the climatic phenomenon called El Niño. For several years, until the United Nations prohibited the practice, drift-net fishing cut heavily into populations.

In the Columbia River system, hydroelectric dams, which chop up young fish in their turbines, have contributed greatly to the problem. Fish in hatcheries, introduced years ago as a way of supplementing wild stocks, have perversely damaged wild populations by competing with them for food and habitat, and by interbreeding with them. Unlike wild fish, biologists say, hatchery fish cannot maintain themselves in the wild without restocking, and interbreeding weakens the survival ability of wild fish as well.

Unlike many other species, salmon die within days of spawning, making the survival of offspring doubly important.

The Upper Rogue watershed, including Little Butte Creek, encapsulates the problem nicely. Little Butte is one of the few remaining major streams containing salmon and steelhead in the Upper Rogue watershed. Much of the rest of the watershed has been closed off to the fish by Lost Creek Dam northeast of here. On their way to the Little Butte, the fish encounter many obstacles. Some wander into irrigation bypasses and end up as carrion in farm fields. Many are lost at dams with inadequate fish ladders.

"It's a tough life," said Mr. MacLeod.

On the Rogue downstream from here, the Savage Rapids Dam frustrates or kills tens of thousands of migrating salmon each year, fisheries biologists say. The National Marine Fisheries Service estimates that removal of the dam would result in the production of nearly 45,000 more salmon and steelhead a year. Mr. MacLeod estimates that production in the Rogue watershed would grow by 20 percent. After a long and contentious battle, Oregon decided late last month to remove the dam in five years.

The ultimate goal is to restore watersheds "from ridgetop to ridgetop," says Mary Lou Soscia, director of Oregon's watershed health program. The future of the enterprise is not free of uncertainties and potential snags. The Oregon state program is financed only through next year and must be renewed. To solidify citizen support, Oregon is relying on a series of locally created watershed councils. One of the more important, in the Illinois River watershed, has been convulsed by what environmentalists see as an effort by irrigation interests to keep them off the board.

There is also a difference of view on how far a restored and protected buffer of vegetation should extend on each side of a stream. The federal plan calls for a uniform zone equal to twice the height of the tallest streamside trees or 300 feet, whichever is greater. The Cummins report points out that this is relevant only in mature conifer forests and not in other kinds of riparian areas. Moreover, the report contends, the widths of setback areas should be adjusted to the size of the stream: One size does not fit all.

"Everybody agrees that there's a considerable need for restoration; what isn't agreed on is the most cost-effective way to do it," said Dr. Daniel B. Botkin, president of the Center for the Study of the Environment and director of the larger study of which Dr. Cummins' work was a part. Further reports are due next month.

Despite the scale of the task ahead, hopeful anticipation abounds. The actual work of restoration "is not very dramatic," said Mr. Babbitt, "but the cumulative impact is enormous, and the payoff is very quick."

For his part, Mr. Grenbemer is anxious to assess that payoff by getting into his snorkeling gear and surveying the population of juvenile salmon produced by the fish now about to move into restored stretches of Little Butte Creek. "The fish counts we take this winter are going to be telltale," he said. "I'm really optimistic."

—WILLIAM K. STEVENS, November 1994

Anglers' Gain Is Loss for Lakes and Streams

California
Winner: brown trout.
Loser: its prey, native
golden trout.

Nevada
Winner: guppies, dumped
by aquarium hobbyists.
Loser: nearly extinct white
river spring fish.

North Dakota
Proposed contender:
introduction of the
zander. Possible loser:
the native walleye.

The Great Lakes
Winner: the fishing industry,
from introduced Pacific salmon.
Loser: native lake trout.

New Jersey
No conflict: plans can-
celed to introduce
Pacific salmon into
Delaware River system.

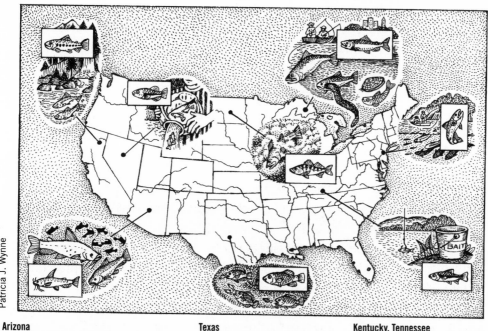

Patricia J. Wynne

Arizona
Winners: several species of trout, catfish.
Losers: native Yaqui catfish and Apache trout, to
hybridization and competition.

Texas
Winner: green sunfish. Loser: its prey,
the native Comanche Springs pupfish.

Kentucky, Tennessee
Winner: southern redbelly dace, escaped fishermen's bait. Loser: native tiny blackside dace.

BIOLOGISTS HAVE LONG CATALOGUED and mourned the havoc that exotic animals can wreak when moved into habitats where native species have evolved few defenses against them. Grim stories abound, from the sharp decline of bluebirds in the United States due to competition from imported English house sparrows and starlings to the devastation of tortoises on the Galapagos by introduced pigs, cats and dogs who prey on the eggs and young.

89

Today, few wildlife biologists would suggest that nature can be improved by introducing exotic species into ecosystems on land. But, some critics now say, such practices remain all too common in lakes and streams, where many state fish and wildlife agencies continue to introduce exotic fish to provide anglers with more exciting sport or tastier quarry. Such widespread practices are a key reason that native fish are the most endangered category of animal in the United States, the critics contend.

Now, biologically displaced fish proliferate in America's waters: Huge chinook salmon from the Pacific Northwest are in Lake Michigan, bass originally from Atlantic drainages are found across the West, Eastern sunfish are in Arizona waters.

Such stocking programs are popular with anglers, whose license fees and special taxes on tackle pay for much of the nation's fisheries management. But some scientists insist that government fisheries managers are stubbornly continuing to make the same mistakes made in terrestrial habitats a century and more ago.

"You could argue that introducing fishes in the United States constitutes one of the greatest ecological experiments in history in terms of size and scale," said David Wilcove, an ecologist with the Environmental Defense Fund (EDF), a conservation group. "But the results of the experiment haven't been so great. The degree of biodiversity loss in all the major river systems in the United States has been astounding."

The pressures on native species occur because fisheries managers often "add big predators to lake or river systems that never had big predators before," Dr. Wilcove said. In some cases, native fishes have been pursued and devoured by the introduced predators. In other cases, introduced species are aggressive foragers that outcompete native fish for food or eat their eggs. In still other cases, native fish species have been genetically swamped by breeding with related introduced fish.

In a 1991 report for the Congressional Office of Technology Assessment, Micheal Bean, an attorney with the EDF, and a widely recognized expert on endangered species law, noted that more than half of the United States fish listed by the Endangered Species Act were threatened by introduced fishes, and nearly one third by species introduced in connection with sport fishing.

Further, a 1989 study published in the journal *Fisheries* by Robert R.

Miller, James D. Williams and Jack E. Williams, all biologists, concluded that more than two thirds of the fish extinctions that had already occurred in the United States were caused at least in part by pressure from introduced species.

According to Dr. Wilcove, the problem has been compounded by introduction of non-game species into American waters, including aquarium species dumped by hobbyists. Small bait fish have been introduced, either unwittingly by anglers or intentionally by government agencies to provide forage for bigger game fish.

More than one third of North America's 850 native fish species could now be considered rare, approaching extinction, or already extinct, according to a report by Dr. Wilcove, Mr. Bean and Philip C. Lee, a biologist at Simon Fraser University in British Columbia, published in September in the proceedings of the North American Wildlife and Natural Resources 1992 conference. That compares with about 11 percent of the continent's breeding bird species and 13 percent of its mammals.

Dr. Wilcove and Mr. Bean pointed out that there are other major pressures that threaten extinction, including pollution and damming. However, they noted, the harm from introductions "is harm nonetheless, and in many cases, easily preventable harm."

Some state agency managers disagree that introductions present a serious problem, suggesting that through careful management, introduced species constitute a sort of engineered improvement to aquatic nature.

"There is an attitude that I would categorize as a purist attitude that we shouldn't do anything that's not part of the natural environment," said Lloyd A. Jones, commissioner of the North Dakota Game and Fish Department. "Others of us have a different approach: that we need to analyze natural conditions and see if we can improve upon them."

Indeed, the practice of manipulating ecosystems to provide better fishing is deeply imbedded in state natural resource agencies, in large part because the sport fishermen who buy licenses expect it.

Dr. Walter Courtenay Jr., a professor of zoology at Florida Atlantic University in Boca Raton, and an expert on fish introductions, said: "The state agencies feel that their responsibility is to satisfy the sportsman, because that's the side of the bread their politics is buttered on. When ecologists try to bring this issue up, the agency people say, 'Don't bother me with conservation of all of our native fishes.' That's a very unfortunate attitude."

Probably the most famous example of damage from intentional introductions is one of the oldest. In the early 1830's, carp were first introduced into the Hudson River. Though carp were once popular game fish in England, and 19th-century proponents hoped for a similar reception here, carp have been generally reviled both as sport and as meat in the United States.

Nevertheless, carp have prospered, particularly in the Midwest, at the expense of native ecosystems. "Carp have a habit of uprooting native vegetation and dirtying up the water," Dr. Courtenay said. "Any fish that prefers clear water simply won't stay there."

The blue tilapia, an African fish, has been introduced for weed control into the lower Colorado River, and now, because it consumes so much of the available food, prospers at the expense of native species.

Meanwhile, aggressive largemouth bass introduced into the Colorado have threatened several native species, including the razorback sucker.

In Nevada, guppies dumped by aquarium hobbyists into thermal springs south of Elko have similarly prospered, but have pushed native species like the white river springfish to the verge of extinction.

In California, the golden trout is at risk because of predation by introduced brown trout in some streams.

In Maine, introduced largemouth and smallmouth bass appear to be suppressing survival of native Atlantic salmon in streams, according to Edward Baum, director of the Maine Atlantic Sea Run Salmon Commission. Mr. Baum noted that for anglers it was "sort of a two-edged sword."

"They have bass," he said. "But as a result, they don't have salmon."

Because of the still inadequately understood nature of aquatic ecosystems, the effects of introductions often are unpredictable. Critics suggest that it may make far more ecological sense to re-focus state fisheries programs on promoting native ecosystems. But not all introductions have been detrimental, argues Paul Brouha, executive director of the American Fisheries Society, which represents fisheries scientists and managers. "To say that all introductions are bad is to ignore the fact that we already have a lot of altered habitats," he said. "Look at the large reservoirs we now have on a lot of river systems. If there were no introductions, they might not have much of a fishery because native fishes aren't adapted to living in a reservoir system."

Dr. Wilcove suggests that, philosophically, fisheries managers may be repeating the hubris of decades past. "In some respects, this resembles the way we treated terrestrial wildlife a century ago: Get rid of wolves and bears and any animals you don't like. Build up populations of game animals like deer. And try to design the terrestrial fauna of the continent to suit your needs. We realize now that doesn't work for terrestrial ecosystems if we want to protect diversity. But in aquatic ecosystems, we're pursuing similar goals."

Part of the reason that issue has gone largely unnoticed among conservation groups and the public at large may have to do with perceptions of what constitutes damage to a waterway, said Dr. Wilcove. "Ecologically, you can alter an aquatic ecosystem as dramatically as converting a forest into a shopping plaza. But as long as there's water flowing, and the water looks clean, most of us would never suspect there was a problem."

—JON R. LUOMA, November 1992

Wild trout feed cautiously and guard their feeding positions jealously. Hatchery trout are more reckless and disrupt the social order.

Wild trout spread their fins and send other body signals to warn interlopers away. Hatchery trout, unaccustomed to the game, tend to ignore the signals. Exhausting fights often ensue, and hatchery trout sometimes evict wild ones from their feeding spots.

The experiment found that after two years, there were fewer wild trout. Although hatchery-raised fish were fatter than the svelte wild fish to start with, they fed inefficiently, expended too much energy, grew thinner and often died.

Glenn Wolff

Hatched and Wild Fish: A Clash of Cultures

TO MILLIONS OF TRUE BELIEVERS, there is nothing more beautiful in all of nature, nothing to make the heart beat faster, than the aristocratic trout and its royal cousin, the salmon. Brilliantly spotted, pink-flanked or simply and elegantly silver, they linger in the mind's eye as paragons of sleek grace and primitive power.

In pursuit of that vision, and to replenish commercial salmon stocks, fisheries biologists over the last half century or so have released billions of hatchery-reared fish into American streams, rivers and lakes. For years, no one thought much about the ecological and genetic consequences of turning them loose.

But now it is clear that fish-stocking programs have transformed the nation's trout and salmon population and may even be threatening the long-term survival of wild fish.

When adult hatchery trout are suddenly thrust into a stream where wild trout have already established a stable social order, "they run around like a motorcycle gang, making trouble wherever they go," says Dr. Robert A. Bachman, a behavioral ecologist who directs Maryland's freshwater fisheries division. The new arrivals charge about the stream in a tight school, something the wild fish would never do, provoking fights everywhere. The conflict and chaos, Dr. Bachman has found, eventually result in fewer fish of either kind.

Other studies have also found that stocking tends to reduce the number of wild trout. The hatchery trout dwindle, too, since they are generally more easily caught and less adept at feeding on wild fare. The outcome is often an impoverished fishery dependent on periodic fixes of stocked fish.

Of more serious concern are the genetic risks posed by stocking programs. The genetic integrity of some wild strains, and at least one species, is being threatened by interbreeding and hybridization. Meanwhile, hatcheries in some cases have produced populations of trout and salmon with less genetic variety than is found in the wild. As these fish breed with wild trout, scientists say, they erode the natural gene pool and may impair the ability of wild fish to adapt genetically to environmental changes.

Awareness of these dangers is encouraging fisheries biologists to preserve and bolster populations of wild fish and is prompting a shift in fishery practices. Some states have imposed strict limits on killing trout, thus limiting the need for restocking. Hatcheries and "put-and-take" stocking programs, in which adult fish are planted in streams only to be caught almost immediately, are being re-examined and assigned a reduced role in many places.

In one sense, stocking from state, federal and private hatcheries has enriched the nation's fisheries, giving millions of anglers the chance to go after trout and other game fish. Nearly 40 million Americans, about 12 million of them trout and salmon anglers, spend more than $20 billion a year on freshwater sport fishing. The widespread introductions have also helped put a delectable and healthful food on many tables.

The ranges of the major species of stream-dwelling trout have been greatly expanded. Brown trout originally were found only in Europe; rainbow trout, in western North America; brook trout, in eastern North America. Now all are established in cold waters across North America—in many cases crowding out the original denizens. Other fish, especially bass, have been widely propagated as well. But trout and salmon account for most hatchery and stocking activity.

In the typical trout or salmon hatchery, scientists say, fish are reared under conditions that cause them to act differently from wild fish. They grow up in concrete tanks where they are usually segregated by size class, in dense concentrations, under unnatural light and temperature. They eat "fish chow," specially prepared pellets of fish meal and other ingredients that resemble dry pet food, and grow used to the humans who cast the pellets into the water.

Under these conditions, fish that rush to the food fastest, disregarding the presence of humans, survive and prosper. In the wild, survival depends on just the opposite response. Besides avoiding fishermen and other predators, wild fish in streams must capture the insects and crustaceans they feed on while expending as little energy as possible in fighting the current. Positioning becomes critical. A fish that uses more energy than it takes in will waste away and die.

When brown trout raised in a hatchery were placed in a stream with wild brown trout, Dr. Bachman found in a study in Pennsylvania, they would "throw caution to the winds," rushing around in search of food. But they spotted wild food less skillfully and swam farther than wild fish to get it. Their energy equations did not balance and they tended to get thin and die.

While they lived, they thoroughly disrupted the ecology of the stream. Wild trout jealously guarded their prized feeding and resting stations. But because hatchery trout do not easily recognize body-language signals used by wild trout to warn away interlopers, they readily antagonized the established residents. Exhausting fights ensued, and the wild trout were often ousted from their preferred spots, disrupting their feeding patterns. The upshot, said Dr. Bachman, was that after two years the stream contained fewer trout, both hatchery and wild, than there were wild trout when the experiment started.

Trout from genetically different local strains, subspecies and even species often interbreed after fish are introduced from one range to another. A dramatic example concerns the rainbow trout and the cutthroat trout, both native to the northwestern United States. In Montana, one of the nation's trout-fishing meccas, the commonest fish is now a rainbow-cutthroat hybrid, said Dr. Robb Leary, a fisheries geneticist at the University of Montana. This hybridization, he said, is probably the main cause of widespread loss of the native cutthroat population. "That's genetic extinction right there," he said.

Regional authorities in the Pacific Northwest are undertaking a new program in which hatchery salmon will supplement wild populations that are declining because of overfishing and habitat loss. As part of an attempt to avoid inadvertent genetic damage, four kinds of genetic risk posed by hatchery operations have recently been identified by a scientific panel of the Northwest Power Planning Council. This is an organization established by Congress to protect wildlife in the region. These are the risks:

- Local extinction of wild fish populations. This can happen when a declining population is reduced even further by the need to obtain wild fish whose eggs can be used in hatcheries. "The hatchery can increase the risk of extinction if you're continually mining the wild parents and if the hatchery fish don't do well and don't contribute to the wild population," said Dr. Anne Kapuscinski, a fisheries geneticist at the University of Minnesota who heads the team of scientists examining the problem.

- Loss of genetic variability. Some important genes can be lost as a result of hatchery operations if, for example, the operators rely on too few parent fish for eggs or if sperm comes from too few males.

- Loss of population identity. This can happen if hatchery fish whose parents came from one stream are introduced into another stream with a different environment. Because of the environmental difference, the local fish populations will have developed different genetic adaptations. The introduced fish may not perform well in the new river. But they will interbreed with the natives, and the resulting hybrids may not perform well, either.

- Domestication of hatchery fish. Hatcheries inadvertently select for characteristics that are inappropriate in the wild. They may also promote an unrepresentative section of the wild gene pool, for example by taking brood stock from fish that spawn just in the first part of a weeks-long spawning run. If the early spawners then predominate in the wild, the population may be less able to survive a poorly timed spell of bad weather or flooding.

Genetic changes become "more of a problem over a long period of time as you increase the number of hatchery fish that are surviving and returning to reproduce," said Dr. Harold Kincaid, a research geneticist at the National Fisheries Research and Development Laboratory, an agency of the Fish and Wildlife Service, at Wellsboro, Pennsylvania. "We gradually lose genetic material," he said, as genomes are "basically broken up" by the modified hatchery population. This, he said, is already happening: "I'm sure it's widespread, no question about that." How serious this will be, he said, remains to be seen.

Alerted to all these dangers, many fisheries biologists have begun thinking wild and changing their practices.

Increasingly, the role of hatcheries and put-and-take stocking is being reduced. A number of states have allowed prime trout water to return to the wild state, with no stocking, while permitting anglers to keep one or two fish a day, or none. In Maryland, for instance, this type of fishing has been expanded. In Maryland streams where natural reproduction is insufficient but the habitat is otherwise favorable, hatchery trout are introduced as small "fingerling" fish and allowed to grow up essentially wild. Put-and-take angling for adult hatchery-raised trout is being restricted to waters that for much of the year are too warm for trout to survive.

"By and large," said Dr. Bachman, "what you're seeing across the country" is a recognition that "where one can manage streams for wild trout, you're better off doing so."

In the Pacific Northwest salmon fishery, commercial fishing will continue to make some stocking necessary. But in a new approach, stocking is considered strictly supplementary and the hatcheries are managed to minimize genetic differences with the wild fish. "It's a pretty hot topic out here; all the states are going into it in a big way," said Dr. Craig Busack, a geneticist with the Department of Fisheries in Washington state. Dr. Busack was one of the first to delineate the genetic threats posed by hatcheries.

One way to reduce the mismatch between wild and hatchery fish is to make hatcheries more natural. There are some precedents for this. At the Connetquot River State Park Preserve on Long Island, which contains a surprisingly pristine spring-fed trout stream, trout are bred in a hatchery section of the stream itself. The trout are screened from human contact as much as possible. As a result, "our fish swim away from you," said Gilbert Bergen, the park manager for the environment. "At every other hatchery they come and crowd at your feet. We're trying to raise these fish as close to natural fish as possible."

That may or may not become widespread, given the large investment in traditional hatcheries. But more and more fisheries experts are convinced that going wild, with hatcheries secondary, is the wave of the future. "It has to be," said Dr. Kincaid.

—WILLIAM K. STEVENS, July 1991

4

OCEANIC
FISH

I f tropical forests had been explored solely by helicopter, with nets or scoops winched down to take samples from the forest floor, it is unlikely that many jaguars or other interesting animals would have been discovered.

Methods for exploring the ocean bottom have been scarcely more thorough. Because little life was found there, biologists assumed that these regions were comparatively barren. Theory supported this view. Following the terrestrial model, that new species are created by geographical barriers, biologists inferred that because the ocean floor is fairly uniform it would not harbor many species.

Better methods of sampling the deep-ocean floor have destroyed the ocean-as-desert theory. Biologists now recognize that an immense diversity of species inhabits the deep-ocean floor. For the biology of these animals to become as well understood as that of terrestrial fauna will be the work of decades.

Meanwhile the ocean's dark abysses and barely explored frontiers surely hold many undiscovered wonders, such as the furtive giant squid and maybe other monsters of the deep.

The World's Deep, Cold Sea Floors Harbor a Riotous Diversity of Life

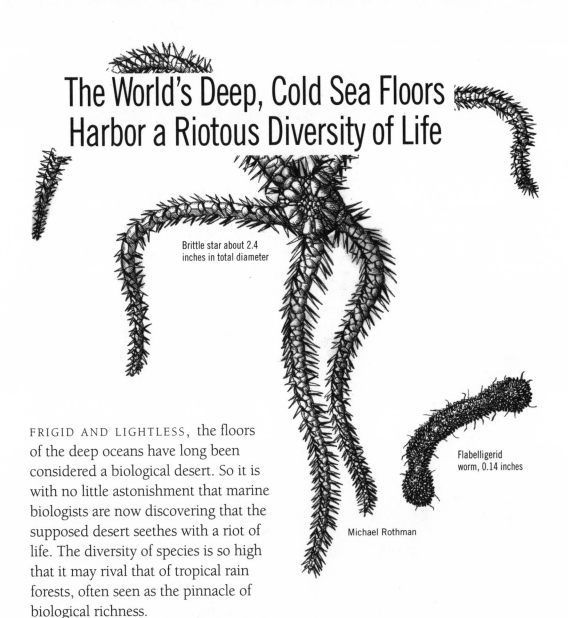

Brittle star about 2.4 inches in total diameter

Flabelligerid worm, 0.14 inches

Michael Rothman

FRIGID AND LIGHTLESS, the floors of the deep oceans have long been considered a biological desert. So it is with no little astonishment that marine biologists are now discovering that the supposed desert seethes with a riot of life. The diversity of species is so high that it may rival that of tropical rain forests, often seen as the pinnacle of biological richness.

The profusion of species on the ocean floor also poses a severe challenge to current theory, since new species are thought usually to require some kind of environmental barrier in order to diverge, like a mountain range, while the ocean floor is more uniform than almost anywhere on land.

Rough estimates for the number of species on the deep-sea floor have now soared to 10 million or even 100 million, hundreds of times larger than the old projections of 200,000 species for all types of marine life.

The new figures also contrast starkly with the sum of the earth's plants, animals and microbes that scientists have so far named, about 1.4 million

103

Tiny haploniscid,
less than one-
tenth of an inch
long

Polychaete worm,
1.4 inches

Mass of
foraminifera,
one-tenth inch
in diameter

A 32-foot-long
lophenteropneust,
a deep-sea worm

species in all. And they match the 10 million to 100 million that experts had projected as possible totals for the number of terrestrial species.

"It's changing our whole view about biodiversity," said Dr. P. John D. Lambshead, a marine biologist at the Natural History Museum in London who studies the abundance of deep-ocean species.

"The quantity of life we've found is incredible," he added in an interview. "All sorts of ecologic theories that looked good, based on terrestrial models, suddenly fall apart. We're having to change all our ideas."

The new creatures of the deep are anything but cuddly or cute, menacing or sinister. Dwelling on or in seabed ooze and often smaller than an aspirin tablet, they include tiny slugs, snails, crabs, bristle worms, ribbon worms, lamp shells, tusk shells, sea anemones, brittle stars and sea cucumbers. The biggest are seldom longer than a banana.

Often miles deep, thriving in pitch darkness under enormous pressure, the mobs of marine invertebrates have now been found in hundreds of deep samples from the northeast and northwest Atlantic, the eastern and western Pacific, and other parts of the global sea.

The variety of life is so high that there is very little overlap among species from various sampling sites, even when they are relatively close together. It is almost as if the animals in any given sample were mostly endemic, that is, species that live nowhere else, as is often found on Pacific and Caribbean isles.

In this case, however, the endemicity is occurring in water—a medium famous for its lack of isolating barriers and its propensity to aid animal migration. Moreover, it is apparently occurring over much of the domain of the deep sea, a dark world that envelops nearly two thirds of the earth.

Though small and ugly by human standards, the newly recognized creatures are considered important because of their possible commercial value, because of their role in maintaining the earth's ecological balance and because of the intellectual challenge of understanding their place in the planet's evolutionary history.

The potential commercial value of the new organisms lies in their great genetic diversity. In general, all kinds of creatures with strange metabolisms from odd places around the earth are starting to be aggressively investigated as possible sources of biological wealth. The hope is to use their exotic genes to develop new drugs, catalysts and agents that can break down toxic wastes.

The discovery seems to give some indirect credence to speculation about the existence of much larger sea creatures that remain to be discovered. If there are krakens, leviathans or other unknown monsters that prey at the top of the rich food chain of the deep ocean floor, they are certainly too big for the kind of small traps so far used in sampling programs.

Not surprisingly, the discovery of the sea floor's biodiversity has set off debates as scientists struggle to understand the unexpected opulence of a supposedly barren world.

"Nobody has explained this," said Dr. Robert R. Hessler, a pioneer of deep biodiversity who works at the Scripps Institution of Oceanography in La Jolla, California. "Everybody comes up with wonderfully plausible ideas. But nobody really knows why you get all these species. The issue is just hanging there."

Dr. J. Frederick Grassle, director of the Institute of Marine and Coastal Sciences at Rutgers University in New Brunswick, New

One-inch-long tanaid

Sipunculid worm, about an inch long

Tripod fish, reaches almost a foot

Jersey, and a leading figure in the field, said the mystery had important implications for understanding the fate of the earth.

"Species diversity is one of the most sensitive indicators of change," Dr. Grassle said. "A lot of highly diverse areas need urgently to be studied because they're disappearing, the rain forests and coral reefs.

"We don't know how threatened the deep sea is," he added. "But in the long term there are going to be changes. So there is some urgency in knowing what's out there."

Scientific theories of life are often rooted in the ideas of Charles Darwin, whose *Origin of Species,* published in 1859, said evolution was partly driven by reproductive isolation. Species often arise, he held, when barriers like mountains or deserts prevent the interbreeding of populations.

In time, groups that become isolated drift apart genetically and physically to form new species, meaning that they are so dissimilar that they cannot successfully procreate.

Land is full of such barriers, both geographic and climatic. But the sea has few—a fact Darwin and his scientific heirs often pointed to in explaining why the land appeared to be so much richer biologically than the sea. This logic seemed reinforced in considering the deep, which not only had few environmental barriers but lacked primary producers such as plants. For food, its inhabitants mainly had to rely on a rain of organic scraps falling from far above or to prey on one another.

Expeditions over the decades that dropped lines and dredges into the deep seemed to confirm the wasteland idea. The few glimmerings of life that were discovered tended to be monotonously similar. The sea cucumbers of the deep Atlantic were virtually indistinguishable from those of the deep Pacific, as many a weary researcher observed.

The first hint that things were radically different came in the late 1960's when Dr. Hessler and Dr. Howard L. Sanders, both then at the Woods Hole Oceanographic Institution on Cape Cod, developed new kinds of bottom-sampling sleds that revealed an astonishing richness in the depths of the north Atlantic.

The breakthrough was simple. The sampling nets that had been regularly towed behind such sleds were replaced with ones in which the nylon meshes were much finer. The new nets caught smaller creatures, and caught them in prodigious numbers. One sampling run hauled up 365 species.

Though startling, the work was slow to be duplicated elsewhere because deep research was so difficult and costly. Moreover, collected specimens were often hard to identify because so few biologists were trained in deep-sea taxonomy. In short, the richness was debatable.

The work was slowly extended in the 1970's to many new sites in the Pacific and Atlantic, with similar startling results. Even so, skepticism continued in some circles because the sampling was imprecise. Sled runs for different times and speeds produced different results. And it was hard to know how far the sleds traveled across on the bottom, a fact that made the density of sampled life ambiguous.

So Dr. Hessler, after he moved to Scripps, worked with a colleague there to develop a device known as a box corer. Like a giant square cookie cutter 20 inches on a side, it was dropped on a line from a ship and cut into a precise volume of muddy sea floor. A seal drawn across the corer's bottom kept the sample from falling out during retrieval.

The box corer worked a revolution in the field, allowing a new level of precision. Now, for the first time, the distribution of deep fauna could be exactly mapped. Though individual samples were small, repeated ones over a region could give a clear reading of species density.

A half dozen sites were studied with such methods in the 1970's and 1980's, with tantalizing results. But the field really developed only after Dr. Grassle, then at Woods Hole, and several other scientists embarked on an extensive study off the East Coast of the United States for the Interior Department's Minerals Management Service, which was considering oil and gas development in deep water.

Armed with a few million dollars, Dr. Grassle, Dr. Nancy J. Maciolek, Dr. James A. Blake and Dr. Brigitte Hilbig, among others, in the mid-1980's dropped box corers measuring one foot square into waters off Delaware, New Jersey, New England, and North and South Carolina. A total of 556 box core samples were taken at sites up to 2.2 miles deep. The feast of life extracted from the muck was so great that taxonomists spent several years identifying all the different types of animals.

"Our results, from the first extensive quantitative sampling of deep-sea communities, indicate a much greater diversity of species in the deep sea than previously thought," Dr. Grassle and Dr. Maciolek wrote in the February 1992 issue of *The American Naturalist,* a scientific journal.

From 272,009 individuals captured by the box corers, the scientists identified a total of 1,597 species. More important, the rate at which new species were added remained high throughout the sampling—in other words, the diversity of life was so great that newness was found wherever a box corer hit bottom. Every square foot of ooze disclosed another dozen or creatures that were unknown to science.

"The number of species continued to rise steadily as more samples and more individuals were collected," the scientists wrote.

Based on the rate of additions, the scientists estimated that the deep sea in general might hold 100 million species of small invertebrates. Assuming that abyssal regions far from continental shelves supported less life, they said, a more realistic number was 10 million species. "This estimate is probably conservative," they added.

It nonetheless provoked strong debate. Dr. Robert M. May, a zoologist at Oxford University, faulted the figures as unsupportable and said that the deep total was unlikely to exceed a half million species.

By contrast, Dr. Gary C. B. Poore and Dr. George D. F. Wilson, Australian biologists, said their own field studies in the Pacific suggested that global species richness was even greater than 10 million.

"We suspect new estimates could be much higher," they wrote in the February 18, 1993, issue of the journal *Nature*.

Other experts, such as Dr. Lambshead of the Natural History Museum, formerly the British Museum, suggested that the estimates would easily rise into the range of 100 million species if the count included even smaller creatures such as thread worms, copepods and ostracods, uncounted hordes of tiny multicellular animals that flourish in the deep ooze.

Dr. Hessler of Scripps, the deep biodiversity pioneer, said in an interview that marine biologists needed to redouble their research instead of their rough estimates. "What we don't know is the rate of species replacement" across the deep beyond the few areas that have been sampled, he said. "That's the big question."

Experts also want more investigations of the riddle behind the diversity—how the deep is able to support such richness, seemingly in defiance of Darwin. Dr. Grassle of Rutgers said the disparity is probably more apparent than real. His work suggests that extraordinarily fine but nonetheless formidable barriers arise in the deep as, for instance, food resources raining down from above collect on the seabed in transient patches.

Another conjecture is that the extra billion years or so that life has been evolving in the sea compared with land may be a factor in the unexpected biological richness of its deep recesses.

Given the vast dimensions of the emerging field, said Dr. Lambshead, conservationists were wrong to focus so exclusively on land ecosystems. "You'll still read in textbooks that 80 percent of all species are in tropical rain forests," he said. "That's rubbish. It simply means that 80 percent of all biodiversity scientists work in rain forests."

He said deep taxonomists are so few, and the new population estimates so large, that just identifying the inhabitants of the abyss could take thousands of years.

"The kinds of numbers we're coming up with are frightening," he said. "If we're only halfway right, many species could be forced into extinction before they're ever described."

—WILLIAM J. BROAD, October 1995

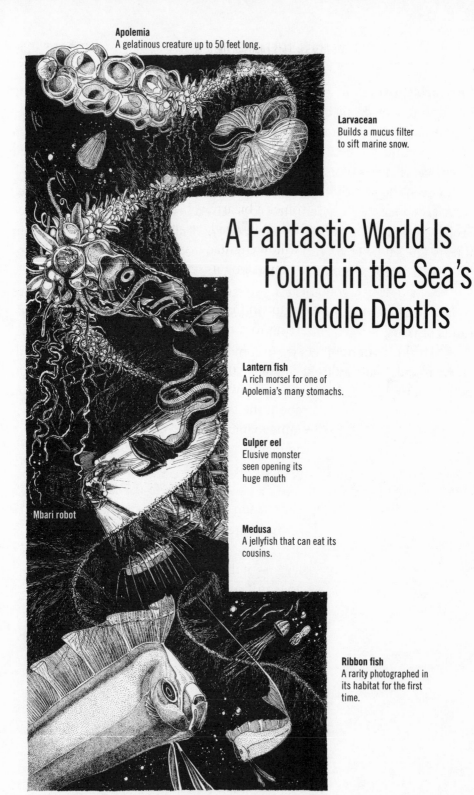

Apolemia
A gelatinous creature up to 50 feet long.

Larvacean
Builds a mucus filter
to sift marine snow.

A Fantastic World Is
Found in the Sea's
Middle Depths

Lantern fish
A rich morsel for one of
Apolemia's many stomachs.

Gulper eel
Elusive monster
seen opening its
huge mouth

Mbari robot

Medusa
A jellyfish that can eat its
cousins.

Ribbon fish
A rarity photographed in
its habitat for the first
time.

Dimitry Schidlovsky

ABOARD THE *Point Lobos,* the darkened room is lit only by glowing computer screens and ghostly images from a robot more than a half mile below, tethered to this ship by an electronic cord. To the left, the pilot guides the robot forward. To the right, the chief scientist scans his video monitor, searching the ocean depths for unusual forms of life.

Suddenly, the blur of marine snow is cut by a thin line. The image on the screen grows until it becomes a bizarre gelatinous creature, perhaps 15 or 20 feet long. Translucent bells at its head pump slowly to move it forward in a long arc. Hundreds of willowy tentacles hang from its slender body, stingers ready to immobilize prey. Along its length are dozens of stomachs, some fat with digesting kills.

"Look, he's got a little Aegina in his tentacles," says Dr. Bruce H. Robison, the chief scientist, referring to a type of jellyfish. The jelly is slowly being drawn up to be eaten.

For ages, the seas have remained an enigma compared with man's knowledge of life on land. Nets and trawls cast into the ocean have produced relatively minor revelations about the fauna of the middle depths, catching some bony fishes and many unrecognizable blobs of jelly.

More recently, robots and manned submersibles have probed the abyss but customarily go right to the bottom, past eerie flashes of living light visible on the way down.

Now, however, this middle realm is starting to surrender some of its secrets, nowhere more rapidly than here, off the California coast in the deep waters of Monterey Canyon.

Dr. Robison and his colleagues at the Monterey Bay Aquarium Research Institute, a private group based in Pacific Grove, are finding strange new types of animal life at a rate of about a dozen species a year. As important, the scientists are discovering how hundreds of little-known inhabitants of the deep behave and hunt, live and die. This invisible world, it turns out, is surprisingly rich and complex.

Six years of such research by the institute have led Dr. Robison to a startling conclusion. He estimates that science has overlooked a quarter to a half of all the creatures in the sea, including otherworldly ones that man can only begin to imagine.

"If an alien civilization came to look at the dominant life form on the planet, they'd be out looking at midwater creatures," Dr. Robison said during a break in the ship's mess. "In terms of biomass, numbers of individuals, geographical extent—any way you want to slice it—these are the biggest ecological entities on earth. But we know virtually nothing about them."

A potent mix of money, vision, and know-how has put the Monterey institute in the lead of investigating this dark world, and the results are beginning to impress scientists around the world. Moreover, the pace of discovery is likely to increase in the next few years as the institute more than doubles in size, adds new gear and personnel, and begins to broaden its investigations to include the depths of distant oceans.

The rush of exploration is courtesy of the institute's founder and patron, David Packard, the billionaire co-founder of the Hewlett-Packard Company, who so far has spent some $60 million to spur the invention of machinery that can probe the watery unknown.

"They're looking at things nobody has ever seen before," Dr. Kenneth S. Norris, professor emeritus of oceanography at the University of California at Santa Cruz, said in an interview. "It's very exciting. Every day you go out on that boat you see some animal or relationship that's never been seen by man. It's a black world down there and they're lighting it up. Nobody's turned on the lights before."

The sea's middle regions are almost incomprehensible in size. By volume, land is estimated to make up about 0.5 percent of the earth's habitable space. The sea comprises the other 99.5 percent. Its vast preponderance is middle waters—those lying between the bottom areas, with their sea cucumbers and creeping fauna, and the upper sunlit regions, inhabited by plants, phytoplankton and the creatures most familiar to man.

The world's research submersibles have repeatedly passed through this middle kingdom but have seldom paused to investigate its subtleties because scientists traditionally view the bottom as more interesting. The problem is also partly mechanical. Typically, a manned submersible carries two sets of weights that make it heavier than seawater. It drops the first set upon reaching the ocean floor, becoming neutrally buoyant to conduct investigations. It drops the second set to ascend, making it lighter than water. The first weights can be dropped in midwater, but that step is usually taken reluctantly since it inhibits further descent.

Dr. Robison got a taste of something different in 1985 as an oceanographer at the University of California at Santa Barbara. Increasingly frustrated with net hauls of mangled and fragmentary creatures, he and some colleagues borrowed a one-man, oil-industry submersible known as *Deep Rover* to explore the Monterey Canyon. The all-electric vehicle, especially designed for midwater work, allowed him to make a wealth of scientific observations and stunning video footage of the area's deep creatures, including their bioluminescent shimmers and sparkles.

Most of the animals were gelatinous, their bodies perfectly adapted to an alien world having no surfaces, no waves and no sunlight.

Those videos in turn helped inspire Mr. Packard, his marine-biologist children and the Monterey Bay Aquarium, which he had founded the previous year, to set up a research institute devoted to exploring the canyon. Mr. Packard insisted on pushing the limits of undersea robotic technology, in large part by making sure the institute's scientists and engineers worked closely together to develop new gear.

"It's an important job to find out what goes on in the vast volume of the ocean," he said at his California home. "I decided to do it right."

Founded in 1987, the Monterey institute began doing research in 1988 with a robot adapted from an oil-industry model used on offshore platforms and pipelines. In addition to the usual mechanical arm, it was outfitted with sonars, color cameras, chemical sensors and collecting devices, including big jars with movable lids and a suction arm that can capture some smaller kinds of undersea creatures and deposit them in sample bottles.

Over the years, the robot has plunged into the depths more than 770 times for daylong dives, a record for deep exploration. So many dives can be done because deep water lies so close to shore here, reducing transit time and allowing oceanographers to go home every night. The Monterey Canyon is an enormous fissure in California's continental shelf that runs up to two miles deep, an invisible scar in the earth's crust grander than the Grand Canyon. In all likelihood, Monterey Bay is now the best-explored area of deep ocean in the world.

On a recent trip, the 110-foot-long research vessel, *Point Lobos,* left the dock at Moss Landing around 7 A.M. and was ready for deep research by 8:45 A.M., idling over waters nearly a mile deep. No land was visible. The day was cool, the sky overcast. The two-ton robot, the size of a small car, was carefully checked out before being gently lowered by crane into the cold water, where its thrusters moved it away from the ship with surprising alacrity before it dived beneath the waves. An albatross floated nearby, curious at the goings on.

Once out of sight, the robot churned its way into the depths, trailed by a long tether, drawing on the ship's electric power. The robot itself was slightly buoyant, so if all systems failed it would automatically rise to the surface.

Dr. Robison, a deeply tanned, 51-year-old man with a gray beard, sat in "the hole," as the dark control room in the ship's forward spaces is known. Rock music filled the air as Dr. Robison explained the workings of the different computer consoles: one for controlling the robot's movements, one for its lights and cameras, and another for its various chemical sensors. The spacious setup with its comfortable chairs clearly had many advantages over the cramped quarters of a manned submersible, not the least of which were a toilet, mess and well-stocked galley right outside the control room.

As the robot sped downward, its main color camera showed a constant blur of marine snow, the gentle rain of organic debris that includes plankton, molts, plant debris, fecal pellets and dead organisms. This detritus from the upper reaches of the ocean, falling by the ton, is the primary food of the creatures of the deep, supporting a riot of filter-feeders, scavengers and predators.

At a depth of 450 feet, where sunlight penetrates only dimly, the robot came upon a gelatinous creature of great delicacy, the kind that would be unrecognizable goo if hauled to the surface in a net.

"Okay, I'm going to zoom in," said Dr. Robison, and the creature's image grew to fill the monitor's large screen.

The creature, known as a larvacean, had spun a giant fan of fragile mucus above itself that acted as a filter to let only the finer bits of marine snow into its domain. Its inner sanctum was a fist-sized structure that resembled a human brain, only translucent, with the creature dwelling inside, largely out of sight. A delicate flap over this home suddenly started to pulse with rhythmic undulations, moving invisible particles into the creature's mouth.

Dr. Robison said the larvacean would abandon its filter when it became clogged with too much debris, and, if the animal was disturbed, it could also drop the gossamer inner house to flit away and start life anew. The abandoned structures, he said, sink rapidly into the deep and undoubtedly are an important source of food for bottom-dwelling animals.

The robot dived deeper, encountering a group of long, thin, pencil-sized creatures at a depth of about 650 feet. They looked like short lengths of fuzzy rope, with tiny threads hanging down. These animals, Dr. Robison said, were a relatively small type of siphonophore, a colonial animal whose different members worked in unison. Each animal had a cluster of swimming bells serving as a "head" and long elastic tentacles for fishing and drawing prey to waiting stomachs.

"These little guys zipping around in here are *Euphausia pacifica*," he said, touching images of shrimp-like animals, "the krill that they feed on." His studies have shown that the small siphonophore, known as Nanomia, occurs in such abundance here that it eats about a quarter of the local krill population, competing successfully against such big predators as squids, albacore tuna and blue whales.

A circular jellyfish pulsated into view, its long tentacles holding an even smaller jelly. "He's chowing down," Dr. Robison observed.

The robot sped deeper. At a depth of 800 feet a beautiful fist-sized comb jelly could be seen. Rows of fused cilia along its transparent body moved in waves, shimmering and iridescent, pushing the animal forward like a vacuum cleaner to sweep up particulate matter.

One fairly rare animal of the upper reaches that was not spotted was Praya, a colonial siphonophore that can grow to lengths of up to 130 feet, longer than the blue whale, which is often considered the largest animal on earth.

The robot dived ever downward. At a depth of 2,550 feet, a long thin line appeared on the screen, growing in size until it became a large siphonophore known as Apolemia. The creature was 15 or 20 feet long, while its family members can reach a length of 50 feet.

The animal's body was too long to see in detail, so the pilot maneuvered the robot close to provide an intimate tour. The translucent swimming bells at its head pumped away, keeping the colony in a slight state of tension. Tentacles hung down from its cream-colored body in wispy profusion, almost like cobwebs. One group held an Aegina, a small medusa. The tentacles were a yard or so in length, each covered top to bottom with stinging cells.

"I've felt them," Dr. Robison noted wryly.

The animal's many stomachs were strung along its body like beads on a necklace, each one associated with a group of finger-like projections. These help move captured prey into a stomach, Dr. Robison said. Some of the stomachs on the living chain were dark and swollen with captured animals, which can include jellies, worms, shrimps and small fishes. One dark stomach was banana-shaped, perhaps holding a marine worm.

In general, Dr. Robison said, the gelatinous animals of the deep were a riddle just starting to be addressed. "Their abundance and obvious importance to the ecology down here has never been assessed," he said, his face lit by the monitor's glow.

The hundreds of expeditions mounted by the institute over the last six years have teased from the sea a number of other startling finds and images. The rare gulper eel has been photographed in its habitat, its huge jaws opening and closing silently. So too, the wispy and enigmatic ribbonfish has been photographed for the first time in its own domain, and scientists were transfixed by the unusual wave-like motions of its long dorsal fin.

The scientists have also studied the canyon walls and floor, discovering areas where nutrient-rich waters seep up through rocks to support vast assemblages of clams and other creatures. They have probed surface waters with computerized buoys and plan dozens of other investigations, including an acoustic observatory to listen for strange sounds in the deep.

"Science is very conservative," Dr. Peter G. Brewer, president of the research institute, said over lunch. "We're trying to crash through the barriers."

Pleased with the results so far, Mr. Packard is expanding the operation to include a new ship ($12 million), a new robot ($3 million), spacious new laboratories and offices ($21 million) and new staff. The number of personnel is to expand from 90 to about 250 in the next several years.

The new robot under development will dive deeper, have more features and be very quiet. Dr. Robison, who is chairman of the research institute's science department, is convinced that the loud noises of the current model scare off some creatures that can hear. The fastidiousness of the silencing effort is evident in the half-built robot. The cooling fan of its computer is held by tiny rubber grommets that damp vibrations.

The new ship under construction, 117 feet long, will have berths and laboratories for 24 researchers and crew members. With a range of 2,500 miles, it will be able to reach Hawaii without refueling, allowing it to go almost anywhere in the world. Dr. Robison is eager to do wide-ranging studies of midwater ecology, comparing Monterey Canyon to the rest of the waters that encircle the planet. The canyon is clearly a rich habitat, but how much richer than the open ocean remains to be seen.

Mr. Packard, 81, is eager to leave a legacy of illuminating this dark realm. "There's a tremendous volume of water," he said, "that has not been sensibly explored."

—WILLIAM J. BROAD, July 1994

A Fish that Can Alter
Its Body Size at Will

SCIENTISTS SAY they have discovered a shape-changing fish that can detect the presence of hungry predators and then turn itself from easy pickings into something much more difficult to swallow.

In a recent report in the journal *Science,* researchers from the University of Lund in Sweden said that when crucian carp, a relative both of goldfish and the common carp, swam in ponds with carp-eating pike, they quickly began bulking up. They piled new muscle growth onto their backs until eventually they became hulking enough to make themselves difficult or impossible for a pike to fit in its mouth.

The fish are the newest in the growing list of animals, including corals, water fleas, barnacles and snails, that have been found to change shape to protect themselves from danger. These animals fortify themselves against both predators and competitors by growing an array of defenses and weapons. Tiny water fleas don helmets and sprout teeth along their necks. Barnacles grow their protective plates in irregular shapes to make themselves more difficult for snails to pry open. And threatened gorgonian corals outfit themselves with newly grown tentacles laden with stinging cells.

But even as the list of creatures that can build such structures has grown, many researchers have been surprised that the newest addition is a fish. The crucian carp is not only the first fish on the list but also the first vertebrate reported to have such abilities.

"It's pretty amazing," said Dr. Roy A. Stein, professor of zoology at Ohio State University. "I find it quite incredible that a fish literally changes its shape in response to the presence of a predator to reduce its susceptibility. When researchers hear about this, they're incredulous. I was a non-believer, too, until I saw their data."

Dr. C. Drew Harvell, associate professor at Cornell University and a specialist in this area, said these so-called inducible defenses had seemed to be largely restricted to plants and to animals that for some reason could not run away. The explanation was that these earth- or rock-bound beasts were forced into protecting themselves by building up their fortresses. "Normally you think of a fish as being mobile and responding behaviorally to its predators," she said, "that it should just swim away."

Dr. Christer Bronmark, associate professor of ecology at the University of Lund and an author of the study, said the changes in the fish's shape were so marked that the bulked up crucian carp were originally mistakenly designated as a different species from those living in ponds without predators. Mistaken identities are familiar for animals with inducible defenses, Dr. Harvell said. The species that she studies, a marine creature known as a bryozoan, can sprout two different structural defenses when threatened; originally it was misidentified as consisting of three different species.

For a hungry pike an undefended crucian carp is the perfect meal; the fish's small, slender body slides down headfirst in one easy gulp. But as with many fish predators, the limiting factor for a pike is how wide or tall its prey is and how far it can stretch its mouth.

As crucian carp begin to grow defensively in a process that can go on for weeks, they soon become too tall to handle easily. Dr. Bronmark says the pike is forced to attack a beefed-up carp from the smaller tail end and then struggle with the fish in its mouth until it finds some way to swallow the meal. Often the fish are too tall to be forced down even when the widest part of the carp's body is entering the widest part of the pike's mouth.

The result is that these oddly shaped carp often manage to swim off unharmed. There's even evidence from another study that once a carp becomes so big that it does not need the extra height it alters its growth so that it is no longer so disproportionately tall.

While inducible defenses are often quite effective, with spines stabbing into an attacker's soft parts and fortified shells too hardened for a predator's claws, in general they are often formed at a heavy cost to their maker. Scientists say the high cost of these structures explains why they are built only when absolutely necessary. Researchers have found that creatures that pour their energy into the rapid production of armaments often fail to produce as many offspring while others die sooner than those who do not have the defenses.

For instance, the researchers from Lund say that the cost of changing shape for a crucian carp is hydrodynamic. In ponds without pike, there are many carp and the main survival pressure appears to be competition for food. After calculating a 32 percent increase in drag on the carp carrying around defensive muscle growth on its back, these researchers deduced that decreased speed in searching for food may be the price the carp must pay.

Dr. Bronmark said he and colleagues are hunting for the cue that alerts the crucian carp to the presence of the pike, though he says it is likely a chemical. Inducible defenses in other animals can be set off by a number of triggers, including chemical cues from the predator, chemical cues from injured prey, as well as direct contact with or damage from the enemy.

Dr. Harvell said that in some ways it really should not be surprising to learn that a fish can change its form. Threats like predators and competitors come and go and researchers are beginning to find that perhaps more useful than any particular kind of defense may be the evolution of plasticity, or the ability to change from situation to situation. In addition to defenses, researchers have begun to find that animals can and do modify any number of structures to suit current conditions, everything from the shape of their jaws to the thickness of their bones.

"Plasticity is the only way for an organism to adapt to a rapidly changing environment within its life span," Dr. Harvell said. "So I'm sure we'll find many, many more examples of this, changing color, changing shape, changing chemistry."

Dr. Stein said: "This opens up the door for more careful examination of plasticity in vertebrates. People are going to have to stop thinking that there's not going to be this kind of change during an individual's lifetime. I know I'm going to look very differently at things."

—CAROL KAESUK YOON, January 1993

The Odds of a
Shark Attack

THE CHANCE OF BEING KILLED by a shark in United States waters, scientists say, is far less than that of dying from a bee sting, a snake bite or even lightning.

But that does not mean there is no risk. Worldwide, there are probably 50 to 75 shark attacks every year, resulting in five to 10 deaths, according to the records in the International Shark Attack File at the Florida Museum of Natural History in Gainesville. About half occur off the United States, where affluence and leisure and hundreds of miles of prime beaches draw people into the ocean.

Once there they often unwittingly find themselves in or near the primary feeding zone of large coastal sharks, often near a dropoff or inshore of a sandbar where a shark may be trapped at low tide. Natural prey tends to congregate in there.

Almost any shark six feet or longer is a potential assailant, but three species have been mainly implicated: the great white, bull shark and tiger. All are found around the world, all are large and all eat large prey like seals or sea turtles.

Some scientists speculate that most attacks on humans—except when a plane crash or the sinking of a ship throws many into the water and causes a shark feeding frenzy—are cases of mistaken identity. Humans are roughly the same size as seals, for instance, and seals are prime prey of the adult great white shark.

Sharks frequently adopt a "bite and spit" tactic. Some scientists believe that a great white, for instance, attacks its prey with a single bite of its formidable teeth, then backs off so as to avoid possible injury in a struggle and waits for the prey to go into shock or bleed to death. Then it moves in and consumes the prey.

Rarely, if ever, has a shark been known to consume a human. The reason may be that the person is rescued before the shark can complete its feeding or because it recognizes that its victim is not normal prey. But if the shark takes a big enough bite, the person may die anyway.

Some scientists believe that the advent of smaller, rounder surfboards with "V" tails since the early 1970's may provoke some California shark attacks. Viewed from below, the silhouette of a surfer paddling on such a board looks very much like that of a swimming seal.

Experts advise swimmers not to wander too far from assistance; to stay in groups, and to stay out of the water if bleeding or menstruating and at twilight and after dark, when sharks are most active and have an overwhelming sensory advantage.

Swimmers are also urged to avoid excessive splashing and shiny jewelry that might look like scales of a prey fish; to avoid uneven tanning and bright-colored clothing, since sharks see contrast very well; to keep pets, with their erratic movements, out of the water; and to avoid areas where fishermen are active and bait fish or other signs of feeding activity are in evidence.

—WILLIAM K. STEVENS, December 1992

Surface waters
Photosynthesis by plants and phytoplankton in sunlit surface waters fuels all marine life, including that of the deep where no light penetrates. From a depth of one kilometer (3,280 feet) down, all light is shut out.

Deep Seas Hide Myriad Beasts

Far below the ocean's surface lies a world without light, without plants, with temperatures near freezing and under many atmospheres of pressure. With the tantalizing glimpses that scientists are getting into this remote habitat once considered to be too inhospitable to support life, they are finding to the contrary that the deep sea teems with strange and unknown creatures.

While submersibles allow scientists to sink to new depths, the few hours' observation they permit frustrate as much as intrigue with hints of what may yet lurk in the abyss.

"Marine snow", the descent of dead plankton, decaying carcasses of whales, seals, fish and other marine animals, and terrestrial wood and leaves from above, provides the only source of nutrients for deep sea animals.

Sea spider

Sea lily

Brachiopod

Atlantic red crab

Soft coral

Octopus

Drawings are not to scale

Sea anemone

Sea cucumber

In Dark Seas, Biologists Sight a Profusion of Life

DEEP IN THE OCEAN, where only the faintest glimmers of light penetrate the frigid waters, there lives a creature that for more than a century has eluded the most diligent scientific pursuers, even though it is believed to be as large as a city bus.

One of the earliest records of the monster, a giant squid, dates to the 1870's when a group of Newfoundland fisherman presented a tentacle, "the horn of a big squid," to the Reverend Dr. Moses Harvey, a biologist from St. John's, Newfoundland, who developed a keen interest in the giants. The "horn," actually the tip of a giant tentacle, had been hacked off the squid with an axe in a battle in which the fishermen were trying to pull it up and the creature seemed to be pulling them down.

Since then, dead and dying squids have washed ashore, but despite all efforts, no one has ever succeeded in seeing the 60-foot-long Architeuthis (pronounced ark-e-TOOTH-iss), meaning "chief squid" in Greek, swimming in its natural habitat far below the ocean's surface.

For marine biologists this elusive squid has become a symbol of how little is known about the creature-filled seas compared with knowledge of life on land.

As scientists chip away at the task of finding the sea's undiscovered creatures, many of which lurk hidden in the depths, they are finding at every turn a surprising abundance of new and previously unknown animals. Some scientists say this great pageant of marine life is so impressive that it puts the meager offerings of the land to shame.

But as intriguing as recent finds have been, biologists who try to explore life on the ocean bottoms continue to be stalled by the difficulties of working there. With their study of ocean creatures restricted by short stays of manned and unmanned submersibles on just a few spots of the ocean floor or the dredging of the ocean bottom with boxes and scoops, scientists find their limited view of deep-sea life growing at an achingly slow pace.

Marine scientists say if the land were studied as spottily as the sea, many of its most impressive creatures, like elephants, anacondas and tigers, would be as elusive as the giant squid.

"It's unlikely that a beast as long as a city bus would escape notice in any terrestrial habitat for long," said Dr. Sylvia Earle, adviser to the administrator of the National Oceanic and Atmospheric Administration, speaking at a conference where researchers discussed marine biology and conservation several weeks ago at Cornell University. "Yet it's been possible for giant squids to elude even highly motivated scientists."

Unlike terrestrial habitats, the seas teem with a seemingly endless array of creatures, some so bizarre that years after their discovery they defy classification even into phyla, the principal groups for related types of life forms. And the deeper the ocean is, marine scientists say, the stranger and more diverse its fauna become. Indeed, perhaps the greatest number of unknown sea creatures waiting to be discovered are lurking in the sea's deepest abyssal plains, a region once thought to be entirely devoid of life.

Dr. J. Frederick Grassle, director of the Institute of Marine and Coastal Sciences at Rutgers University, listed the phyla that he and colleagues recently pulled from the depths off New Jersey and Delaware. Some are familiar, like the Cnidaria, a group that includes jellyfish, anemones and corals, and the Mollusca, the family of snails and clams. But besides these are a multitude of unusual animals like lamp shells, peanut worms, moss animals, ribbon worms, beard worms and many others that lack common names. Dr. Grassle and a colleague, Dr. Nancy Maciolek, of the ocean sciences unit of Battelle Memorial Institute in Duxbury, Massachusetts, published the study in February in *The American Naturalist*.

"You name any kind of odd group that you've ever heard of or seen," Dr. Grassle said, "and they're there in the deep sea."

Dr. Frank Talbot, a marine ecologist and director of the National Museum of Natural History at the Smithsonian Institution, recently sampled the life at the deep-sea bottom. "I found that one drag would bring up glass fibers, which come from the glass rope sponge. You'd have great bundles of this stuff and a whole set of one kind of animals. And the next time you'd go down and you'd come up with the ooze from the bodies of many small skeletons looking like concrete with stones rafted from Antarctica, a big concrete-like mix. If you take one grab after the next, each grab has very little overlap with the one before."

Dr. Grassle said that in their recent study covering an area of the deep sea no bigger than two tennis courts he and colleagues found an abundance of 90,677 individuals representing more than 14 different phyla, a feat impossible to match in any terrestrial habitat. While counts vary slightly from scientist to scientist, there are estimated to be no more than 11 phyla in all terrestrial habitats combined, only one of which, the onychophora, an obscure group of tropical worm-like creatures, is restricted to land. The sea on the other hand is home to 28 phyla, 13 of which are found nowhere else, neither on land nor in fresh water.

Even the tally of 11 terrestrial phyla tends to overstate the land's diversity. Most species belong to just a few of its 11 phyla. The myriad species of insects and spiders all belong to the single giant phylum of the arthropods. All mammals, fish, birds, amphibians and reptiles belong to another single phylum, known to zoologists as the chordates.

"We have been enormously overemphasizing biodiversity on land," said Dr. Elliott A. Norse, chief scientist at the Center for Marine Conservation, "in contrast to the biodiversity where it is, in the sea."

The sea not only has more phyla than the land but scientists are also finding that these phyla may be richer in species. In the same deep-sea study from which Dr. Grassle and colleagues documented an abundance of marine phyla, researchers found many new species as well. The study yielded 798 species, 460 of which had never before been seen. "What we're finding is, our real conclusion is that we can't estimate the total number of species in the deep sea," Dr. Grassle said. "There are just enormous numbers."

In shallower waters, like those over a continental shelf, each new sample of an area brings up fewer new species, he noted. "That tells you you've found most of what's there. In the deep sea you don't get that feeling at all. Every sample seems to bring up something different."

The sea has greater diversity of habitats, too, says Dr. Norse. "The land has a film of life that mostly extends from 100 feet above the ground level and then a few into the ground," he said. "The sea is occupied by living things from its surface all the way down to the bottom of the sea, sometimes as much as 36,000 feet."

The heterogeneity of sea environments, some scientists say, contributes to the great diversity of life in the seas.

Scientists once believed that the ocean floor was rather uniform. Because in the deep sea there is no light for plants to grow and serve as food, the detritus that rains down from above is the only source of nutrients. "Because there isn't a lot of food getting to the bottom, everywhere there is a patch of food, it makes it very different from the surrounding area," Dr. Grassle said. "It's that kind of heterogeneity that allows species to diversify."

But biologists say that the most important reason that the ocean's life forms are so richly diverse may be that it is the birthplace of life. While humans tend to think of the seas as remote and foreign environments, it is actually life on land that is more the exception than the rule.

"There are only a small number of organisms that have evolved the basic tricks of living outside the ocean," Dr. Norse said. "Life on land is really a remarkable series of variations on their few themes."

The greater range of life's variations, even in the dark and near-freezing waters of the deep sea, provides sights so strange that they seem to be of another world. Scientists who sink down to the sea floor in submersibles or who view the deep sea through the eyes of remotely deployed cameras may be assailed by the strangest of sights, like the graceful dance-like movements of the cirrate or hooded octopus, its pink arms swaying in the deep-sea currents, or great herds of sea cucumbers grazing the nutrient-rich sediments as they march in the slow motion so characteristic of this eerie world.

Scientists exploring these deep-sea pastures have rediscovered creatures that were known only from fossils and thought to have become extinct millions of years ago. Some of these living fossils, like the sea lilies, have been placidly passing the eons in the watery darkness since long before the dinosaurs ruled.

The sea lilies were quickly recognized as kin to their closest living relatives, which in fact are not lilies but starfish. But other living fossils like Paleodictyon, first seen in photographs from the mid-Atlantic ridge, seemed so strange to the eyes of researchers that they were assumed to be the result of a photographic problem and not living creatures at all. A Chinese checkerboard of dots, this creature baffled expert biologists and was instead identified by paleontologists. It now shares the name given to similarly patterned life forms seen in sedimentary muds that hardened into rock hundreds of millions of years ago.

Scientists estimate that less than one tenth of 1 percent of the deep sea has been surveyed. Given these estimates, it is perhaps not surprising that a giant squid could defy pursuit for so long. Scientists concede that other creatures, perhaps even larger and stranger than the monstrous Architeuthis, may continue to defy discovery in their vast watery refuges. It is no wonder then that many of the sea's rarer creatures are less likely to be seen in the flesh than on land in the form of fossils millions of years ago. One reason for that is the extreme difficulty biologists face in exploring the deep oceans, and scientists say much work is still conducted in 19th-century style, dredging up the bottom.

"It's analogous to flying over the surface of the land at night with a couple of strong lights and throwing down a sampling device off the airplane and seeing what you get," Dr. Norse said. "It's very hard to put together a picture when that's the way you have to sample."

The deep sea, as remote as it is, has not remained impervious to the human presence. Curious and intrepid scientists who have risked their lives to explore the deepest ocean canyons have sometimes completed their journey only to find at the bottom of the abyss a carpet of cigarette tins, burned coal, beer cans, license plates and other urban refuse.

"There's no doubt that we've managed to clutter up some portions of the deep-sea floor, particularly along shipping lanes," Dr. Talbot said. But he added, "Most of the ocean floor is really pristine."

If so, there is still a chance of protecting the deep oceans before they suffer the same degradation inflicted on tropical forests. Scientists warn that life in the cold, dark regions of the oceans progresses so slowly that any wound will take far longer to heal than in more productive areas like the tropics.

Dr. Earle says interest in biodiversity is continuing to grow, more land-loving scientists are starting to work with the sea, and the sea is yielding more and more of its secrets. This summer another team of scientists will set off in search of Architeuthis. "We'll be looking for the giant squid off Bermuda this summer," said Dr. Clyde Roper, curator of mollusks and a squid expert at the National Museum of Natural History at the Smithsonian Institution.

"I'm ever hopeful that we'll see it, though I think the chances are small. It's a bit like a needle in a haystack because we're not sure exactly where they live. But if we're going to be seeing one, I'd sure like to be around when it happens."

—Carol Kaesuk Yoon, June 1992

The Mystery of Why 25 Fish Species Are Warm-Blooded

Amy Bartlett Wright

Body Heat and Feeding Range
The tuna (top) had large heat-generating muscles and a fast metabolism; a heat-exchange system in veins and arteries keeps heat well within the body, away from gills. Scientists believe that some fish, like the marlin (bottom), developed a similar warming adaptation to extend their feeding range into colder ocean reaches. The marlin has a special heat-generating muscle that warms only its brain and eyes.

THE VAST MAJORITY OF FISH SPECIES are cold-blooded, meaning that their body temperatures fluctuate with the surrounding water. Some 25 species, however, can keep their eyes, brains or entire bodies warm, independent of ambient temperatures, as birds and mammals do.

For years scientists have debated which of two competing theories better explains this finny ability. One was that the fish acquired their warming techniques primarily so they could expand their ranges into colder regions of the ocean, which promised new sources of food. The other hypothesis was that the techniques allowed the fish to increase their aerobic capacity so they could be more active.

Now Dr. Barbara A. Block, an animal physiologist at the University of Chicago, has concluded that new genetic evidence makes a strong case for the first theory, called niche expansion. She discovered that the warming techniques evolved not just once, but on three separate occasions.

"If it had evolved only once, it might not seem important," said Dr. Block, whose study was published last month in the journal *Science*. "But we're getting a clear message that keeping the central nervous system and eyes warm has an advantage."

Scientists have known for more than a century that all species of tuna are warm-blooded, or endothermic, a condition that is rather a feat for a fish. The gills of a fish act as radiators, passing heat out to the surrounding colder water as blood circulates through them to pick up oxygen. Tuna, unlike other fish, have a large, central heat-generating muscle and a high metabolism, and so generate large amounts of heat. To prevent this heat from being lost, they have developed a system of heat exchanges among their veins and arteries that keep heat well inside the body and away from the gills.

In the 1980's, scientists were surprised to find that some other fish species, including the butterfly mackerel and billfish like the marlin and swordfish, are also endothermic, although in a more limited way. These species have a large muscle near the eye that evolved into a specialized heat generator used to warm only the brain and eyes.

To probe the question of why all these species developed warming techniques, Dr. Block examined their evolutionary relationships. In birds and mammals, she explained, endothermy is a primitive trait that occurs in all species. But in fish, it is considered an advanced feature, one that not all species have. "We thought the steps leading to this trait could perhaps be seen if we knew something about the relationships of the fishes," said Dr. Block.

Evolutionary trees are usually based on morphology, the physical similarities and differences found in species. Using that method, however, scientists have never been able to agree on how the tuna, billfish and butterfly mackerel, all scombroids, were related.

Dr. Block built upon the morphological work by examining the similarities and differences found in a gene common to the three species. Using a computer program to scrutinize the DNA of each species, she developed a new diagram of their history. She then located the junctions at which endothermic fish emerged. At each evolutionary branch where endothermic species were found, their closest living relatives were cold-blooded, indicating that the trait had not been inherited from some common endothermic ancestor.

The advantage, as seen in a comparison of the ecology of the endothermic species and their closest cold-blooded relatives, was an expansion of the area where the fish could thrive. The endothermic species dive into colder waters than their close relatives do, presumably to harvest food unavailable elsewhere. Bluefin tuna, for example, feed in polar regions and return to the tropics to breed. Swordfish feed on the cold ocean floor and come to the surface only at night, encountering a 66-degree Fahrenheit temperature change in 30 minutes. Although the body temperatures of cold-blooded species fluctuate with changes in the water temperature, they are not able to withstand such rapid or extreme changes.

"We can see a clear correlation," said Dr. Block. "The endothermic species have broad thermal niches, and the only other thing they have that others don't is the ability to warm their heads."

Dr. Donald Stevens, an expert on endothermic fish at the University of Guelph in Ontario, said Dr. Block had developed "a very strong argument" for the niche expansion theory, but added, "It is not absolutely definitive."

Dr. Block agreed that niche expansion might not account for every evolution of endothermy. She still has questions about why tuna warm their entire bodies, while other species warm only their heads.

"We can't say conclusively that tunas warm their muscles and viscera for the same reason they warm their heads," said John C. Finnerty, a graduate student who worked on the study. "The aerobic capacity theory may still be part or all of the reason why they warm their bodies."

"Endothermy clearly is not necessarily an all-or-nothing condition," said Mr. Finnerty. "It can occur in different stages and in different forms."

—CATHERINE DOLD, MAY 1993

A Burrowing Fish
Shapes the Sea Floor

A SINGLE SPECIES OF TILEFISH, burrowing and excavating in the ocean floor, has played a surprisingly central role in shaping the continental shelf off the East Coast of the United States, researchers have found.

In 41 submarine dives, scientists have discovered that tilefish protecting themselves from sharks dig shelters from Cape Cod to North Carolina. Far from being a smooth accumulation of sediment, as long postulated, much of the ocean floor along the edge of the continental shelf is dominated by the resulting grottos and deep burrows, especially where the shelf is cut by canyons formed when sea level was far lower in the ice ages.

The surprising extent to which one species of fish has altered the sea floor was recorded from the four-man submersible *Johnson-Sea-Link,* operating from the research vessel *Johnson* of the Harbor Branch Foundation of Fort Pierce, Florida.

Tilefish, which reach three feet and 60 pounds, are shaped somewhat like cod but carry a fleshy appendage behind their heads. They are found along the Atlantic seaboard from Nova Scotia to South America and appear unusually sensitive to cold.

They are found chiefly in the region where the continental shelf begins to drop off—at depths of 400 to 900 feet. This habitat, Dr. Churchill B. Grimes of the National Marine Fisheries Service, who is a participant in the study, explained recently, is a warm layer of water sandwiched between layers too cold for the fish's survival. Below is the frigid water of the deep ocean, and above is the shallow coastal water that becomes deeply chilled every winter. The temperature of the intermediate layer where the tilefish live hovers between 48 and 57 degrees Fahrenheit.

According to Dr. Kenneth W. Able of Rutgers University, another participant, tilefish were first described scientifically in 1879 and the Bureau of Commercial Fisheries, the forerunner of the National Marine Fisheries, hoped the fish would become a major food source. But in 1882 billions of tilefish suddenly died. Ships approaching New York crossed a band of floating fish 70 miles wide.

Scientists suspect that cold water invaded their habitat, possibly after a succession of cold winters, and for a time the species was thought to be extinct. In the 1890's, however, the fish reappeared. Ten million pounds were caught in 1915, and since 1972 they have been an important fishery, sold in large quantities to restaurants, largely through the Fulton Fish Market in New York and featured on many restaurant menus.

But submarine explorers over the last few years have seen abandoned burrows with increasing frequency, indicating a decline in tilefish population. Overfishing is believed to have reduced the population by a half to a third from 1979 to 1982.

Dr. Able, in an interview, said he believed tilefish excavated the burrows as shelters from predators, including hammerhead sharks sometimes seen in the area.

The *Johnson-Sea-Link* dives were conducted off the East Coast from 1980 to 1984 and last summer in the Gulf of Mexico. Three types of burrows were observed: cavities under boulders, vertically dug burrows, and horizontal ones in canyon walls.

Alongside Lydonia Canyon and Veatch Canyon southeast and south of Cape Cod the fish burrow under large boulders that were dropped there by melting icebergs in the last ice age.

Farther south, on the gently sloping outer flanks of the Hudson Canyon, an extension of the Hudson River Valley through the continental shelf, the fish dig straight down into the ocean floor. Their vertical burrows were found to dominate 300 square miles on both sides of the canyon. The canyon was formed 10,000 to 13,000 years ago by intense discharge of water when the last ice sheet was melting and low sea level exposed the continental shelf. Because the canyon is so deep, its steep lower walls are too cold for a tilefish habitat.

The scientists estimated the density of the Hudson Canyon burrows to be 1,234 per square kilometer. They are conical, and some, enlarged by crab burrows and cave-ins, are more than seven feet deep with top diameters of as much as 10 to 15 feet. The crabs are believed to be a primary food source for the fish.

To monitor one Hudson Canyon burrow for 24 hours, explorers mounted a 35-millimeter camera and a strobe on a tripod, for photographs every two minutes. The burrow appeared to be the home of a male and female, who foraged for food half the time and remained in the burrow the rest of the time.

The scientists believe vertical burrows are the primary tilefish habitat throughout the southern New England and Middle Atlantic regions.

Other canyons are a favored tilefish habitat because exposed layers of clay in their walls provide ideal places to burrow. Pueblos, homes carved in canyon walls, were first described in 1977 by Dr. John E. Warme of the Colorado School of Mines. More recent surveys have confirmed that they are all dug horizontally into exposed layers of stiff gray clay. They range in size from holes just large enough for entry by one fish to gashes 10 feet long, three feet high and three feet deep.

In their observations, scientists found that multiple openings sometimes led to a single large grotto, recorded by the researchers when they observed a conger eel enter one burrow and emerge from another or when dye injected into one hole reappeared from others.

Clouds of fine sediment were sometimes seen coming from the holes, presumably washed out by swimming activity within and the currents it generates. Scientists believe the fish excavate their dens this way.

The researchers also suspect that the fish excavate with their mouths, because the explorers have seen what appear to be mouthfuls of clay near the burrows.

Tilefish burrows were often weakened around the rims through additional burrowing by crabs, which are very common at these depths. The resulting honeycombs and linked tunnels make the sea floor vulnerable to collapse, producing deep pits.

Such collapses over the thousands of years since the last ice age "may have played an important part in shaping bottom topography around Hudson Canyon," Dr. Able and the other researchers report in the current *Environmental Biology of Fishes*.

The fish seem strongly attached to their burrows. Although those using cavities under glacial boulders made no effort to return when chased away by the submarine, those farther south, on sides of the Baltimore Canyon, refused to move even when prodded with the submarine's manipulator arm.

Some fish proved so loyal to their pueblos in Lydonia Canyon that the same ones were recognized when researchers returned a year later.

Authors of the *Environmental Biology of Fishes* report, in addition to Dr. Able, were Churchill B. Grimes of the National Marine Fisheries Service, and Robert S. Jones of that agency and the Harbor Branch Foundation. Joining them in a report on fish erosion of the sea floor in the *Journal of Sedimentary Petrology* last September was David C. Twichell of the United States Geological Survey.

—WALTER SULLIVAN, July 1986

5

THE SEA'S THREATENED FISH

The oceans are so large and bountiful that it might seem impossible for people to devastate them. Yet devastation abounds.

One fishery after another has collapsed. Having depleted the fish that live near the surface, commercial fishermen are hunting fish of the deep ocean. Because the metabolism of creatures that inhabit the dark, cold waters is much slower, they are also much longer lived. The orange roughy, now being commercially fished in waters off New Zealand, does not reach sexual maturity until it is 30 and can live to be 150 years old.

Few fishing methods are discriminate. Long-line fishing is a snare even to birds, particularly the albatross, that may fly 10,000 miles only to end its life on a bait line.

People's appetite for sushi has sent the bluefin tuna spiraling toward extinction. The shark has met a still more merciless predator in the form of those who fancy its fins.

The ancient coelacanth of the Comoros, having endured for millions of years, is now threatened in its lava cave refuges.

The tragedy of the commons is being played out on the high seas. Many efforts are under way to stem the worst effects of the exploitation, but it is not yet evident that they will succeed.

Creatures of the Deep
Are Hauled to the Table

Dory
Zenopsis ocellata and others
Up to 2 feet

Rattail
Macrourus berglax and
others Up to 3¼ feet

Deep-sea red crab
Geryon trispinosis; size varies

Alexis Seabrook

DWELLING IN DARKNESS, little known or understood, the animals of the deep sea are traditionally thought of as mysterious or monstrous. But that is changing. Increasingly, commercial fishermen are scouring the global deep to catch them by the ton. Creatures of the abyss, frequently grotesque by human standards, are being carefully prepared and marketed to hide their murky origins when sold in stores and restaurants.

Scientists worry that the rush for deep-sea food is beginning to upset the delicate ecology of some of the sunless regions that cover more than half the planet, threatening to tip the evolutionary balance. But chefs and fishermen are eager to probe the new frontier. Deep harvesting, they say, can be prudent and the results quite tasty.

"They're an incredible hit," Michael Stafford, the chef at the Captain Daniel Packer Inn in Mystic, Connecticut, said of the giant deep-sea shrimp he has started to serve. They are known as Stonington Reds and Royal Scarlets and are sometimes as large as lobsters. "They're tender and sweet, and they cook really quickly," he said. "They sell themselves."

141

Other deep creatures now being harvested or targeted as seafood include rattails, skates, squid, red crabs, orange roughy, black oreos, smooth oreos, hoki, blue ling, southern blue whiting, sablefish, black scabbard fish and spiny dogfish.

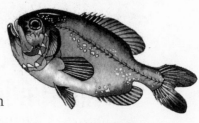

Orange Roughy
Hoplostethus atlanticus
11¾ inches

The search for palatable fare in the sea's depths is partly fueled by fishing wars and the collapse of such shallow fisheries as the Grand Banks off Newfoundland, where the search for wholesome food and short-term profit has driven such popular species as cod and haddock to the verge of commercial extinction. Worldwide, after centuries of steady growth, the total catch of wild fish peaked in 1989 and has subsequently declined.

Spiny dogfish
Squalus acanthias.
Up to 4 feet

Foraging deeper is often a survival strategy for fishermen forced out of traditional grounds by government regulators and pressed by creditors to make payments on expensive boats. In the United States, the investigation of deep fisheries is encouraged by the National Marine Fisheries Service of the National Oceanic and Atmospheric Administration, a federal agency that aids such exploratory work with millions of dollars in grants. The exact amount of the catch from the deep seas is not known.

Royal red shrimp
Pleoticus robustus
Up to 9 inches

But scientists and environmental groups warn that so little is known about deep-sea creatures that some face ruin. Animals that inhabit the icy depths, experts say, often grow and reproduce very slowly, making their populations particularly vulnerable to disruption. Scientists recently discovered that the orange roughy, which dwells up to a mile deep off New Zealand and has been heavily exported to the United States for a decade, reaches sexual maturity in its 30's and lives to an age of 150 years or more.

"Deep fisheries can provide a pulse of good fishing but they're often not sustainable," said Dr. Jack Sobel, a senior scientist at the Center for Marine Conservation, a private group in Washington.

The problem is illustrated by the orange roughy, whose principal stocks around New Zealand recently collapsed. "They never should have been exploited at all," said Mike Hagler, a fisheries expert in Auckland, New Zealand, for Greenpeace International. "People wouldn't eat rhinoceros or any other land creature that they knew was threatened by extinction. But they're eating fish like orange roughy without a clue to what's happening."

Peter J. Auster, science director of the National Undersea Research Center at the University of Connecticut in Avery Point, said knowledge and moderation were essential to avoiding trouble. "Some of these fisheries might be sustainable with a five-boat fleet but would be rapidly depleted if 15 or 20 boats decided to go after them," he said.

"The attraction is understandable. For the individual fisherman trying to pay off the boat or put his kid through school, and facing Draconian rules to reduce pressure on overexploited stocks, this is a wide-open area. It's a place to go and use your wits and knowledge to try to make a living, which is what fishermen have been doing for centuries."

Bruce Morehead, an official of the National Marine Fisheries Service, said his agency was aware of the risks that accompanied deep fishing and was working closely with commercial fishermen to strike a balance between exploitation and conservation. "If you don't accurately assess the stocks," he said, "you can accidentally kill them off."

Although limited commercial fishing of the deep has been practiced for decades around the globe, new sciences and technologies are rapidly making it more practical and efficient, even as its allure grows because of the worldwide depletion of shallow stocks. Hauls that once took days can now be done in minutes.

From the start, the main tool has been a long steel line and a stout net or bag-like trawl, which can be dragged through the midwater or across the dark bottom, its way eased by rollers and wheels. Now ships are bigger and faster, lines are thinner and stronger, and winding drums are wider. Most importantly, the maws of trawls can be enormous, often hundreds of feet wide.

Former military technologies are helping the hunt, including radars that let boats navigate in dense fogs, sonars that locate deep schools and navigation satellites that pinpoint locations so vessels can return to rich locations.

As a result, the bottoms of many continental shelves and shallow seas have been scraped repeatedly in the hunt for deep seafood. Increasingly, the action is moving into the inky depths.

Recently, fishermen have begun pinpointing zones of deep fertility by studying maps based on formerly secret military data, which reveal hidden sea-floor features. For instance, a map made public in October by the National Oceanic and Atmospheric Administration in one step doubled the number of known undersea mountains, which often are associated with upwellings of deep, nutrient-rich water and dense aggregations of sea life.

Deep creatures hauled up as commercial seafood include deep-sea red crabs found off the northeastern United States and southwest Africa, according to Dr. John D. Gage and Dr. Paul A. Tyler, authors of *Deep-Sea Biology* (Cambridge University Press).

In the North Atlantic, Russians pioneered the harvesting of rattails, a ubiquitous deep fish with big eyes and a long tail that swishes back and forth in a sinuous motion. This fishery, which at its peak produced more than 80,000 tons a year, is now in decline. "Whether this has been caused by too-intensive fishing" or natural shifts is unclear, the scientists say.

In the Pacific, the sablefish, or black cod, is fished at depths of up to a mile, and recent research suggests that individuals can live up to 70 years. Anxious over its slow rates of reproduction, Dr. Gage and Dr. Tyler warn that "the stock could become quickly depleted" unless deep fishing is carefully regulated.

A bestseller from the depths has been orange roughy, a pug-nosed warrior that feeds on prawns, squid and fish, living on deep banks and around sea mounts near New Zealand. Trawlers in their heyday could sometimes catch 50 tons an hour. But after a gold rush that resulted in the export of hundreds of millions of dollars a year in orange roughy, the New Zealand fishery recently collapsed to a tiny fraction of its former size.

Now the industry is seeking to develop fisheries for alternative deep species such as black oreos and smooth oreos.

"In deep water you get high costs, high technology, high finance, and consequently, the pressure is on to heavily exploit the stocks," said Mr. Hagler of Greenpeace International. "You're up against the wall, which pushes fisheries beyond the limits of economic sustainability."

Scientists say the collapse of deep stocks such as orange roughy can have all kinds of hidden implications because they are apparently preyed on by such large animals as sperm whales and giant squids. Their disappearance, experts warn, could disrupt delicate food chains.

Off the United States, the National Marine Fisheries Service is helping industry explore fisheries for deep shrimp, rattails, chimaeras, orange roughy, smoothheads, slackjaw eels, blue hake, skates and dogfish, which the National Fisheries Institute, an industry group, in an effort to improve their marketability, has renamed cape shark.

"We're excited because these may open up an avenue to try something new and relieve the pressure on cod, haddock, some of the flounders and other species that are depleted at this point," said Kenneth L. Beal, a manager in Gloucester, Massachusetts, with the northeast regional office of the National Marine Fisheries Service.

One pioneer deep shrimper is Capt. Bill Bomster, who with his three sons runs the 100-foot *Patty Jo* out of Stonington, Connecticut, and sells his catch to the Daniel Packer Inn in nearby Mystic. He has plied the edge of the continental shelf from Block Island to Virginia, dragging the bottom with a 200-foot-long trawl whose mouth is three feet high and 150 feet wide. The trawling is done at depths of up to a half mile.

"What's nice is that the water is very cold and exceptionally clean," he said. "That makes the catch far superior to gulf or farm-raised or any other shrimp. Some of the damn things are 10 inches long."

As stocks of better-known fish shrink and international quotas tighten, experts say the deep will increasingly be targeted as a source of seafood, which is fine with Mr. Stafford at the Daniel Packer Inn.

One of his deliveries of deep shrimp was accompanied by a strange kind of crawfish that nobody could identify, even his friends who are marine biologists.

"One of the arms was really long," Mr. Stafford recalled, adding that the exotic crustacean went into the pot along with the big shrimp.

"It was really tasty," he said.

—WILLIAM J. BROAD, December 1995

Debate Erupts over Peril Facing Ocean Species

FOR THE FIRST TIME, some scientists are saying that some species of ocean fish and invertebrates are reaching the perilously low levels where extinction becomes a real possibility. Until now, only a handful of marine organisms have been scrutinized and placed on the red list along with thousands of ter-

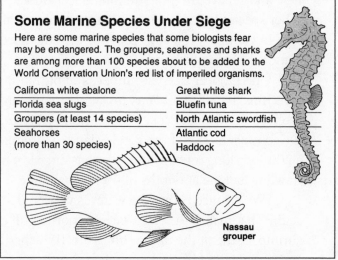

Some Marine Species Under Siege

Here are some marine species that some biologists fear may be endangered. The groupers, seahorses and sharks are among more than 100 species about to be added to the World Conservation Union's red list of imperiled organisms.

California white abalone	Great white shark
Florida sea slugs	Bluefin tuna
Groupers (at least 14 species)	North Atlantic swordfish
Seahorses (more than 30 species)	Atlantic cod
	Haddock

Nassau grouper

Baden Copeland

restrial and freshwater creatures around the world. While it has no legal force, the list is used as a guideline by policy makers. The organization that maintains it, called the International Union for the Conservation of Nature, is based in Gland, Switzerland, and has many governments among its members.

The oceans have had their share of environmental trouble, of course. The plight of whales, other marine mammals and sea turtles is a familiar tale of ecological woe, and loud alarms have also been raised about the rapid depletion of many commercial stocks of ocean fish. But depletion is not endangerment, and it has long been assumed that the sea is so vast and fecund that marine fish and invertebrates are generally in no danger of extinction at the hands of humans.

Not necessarily so, say the marine biologists who cite growing evidence that many oceanic species may be just as imperiled as their terrestrial counterparts, and largely for the same two reasons: the overexploitation of long-lived species that cannot reproduce fast enough and the disruption or destruction of narrow habitats to which many species are confined.

Ten years ago, or even five, "it was inconceivable that endangerment could occur in the ocean," said Dr. Gene Huntsman, a longtime fisheries biologist who recently retired from the National Marine Fisheries Service. Today, many scientists concur with Dr. Huntsman when he says, "I believe we have created true endangerment for some species."

The roster of imperilment would grow, many biologists say, if marine invertebrates not yet considered for the red list were to be added to it. If the roster were to be so expanded, its members would range from the humble and obscure to the spectacular and renowed, including these:

- The delectable white abalone of California, its population so depleted by fishing that little or no reproduction is taking place and extinction appears likely.

- Florida sea slugs, their beauty belying their name and their like found nowhere else, apparently on the ropes because of coastal development.

- A whole family of groupers, including at least 14 species, that are sitting ducks for fishermen because they never leave the shallow-water patches of coral reef on which they live.

- Sea horses, confined to grass beds around the world where they are easy prey for suppliers of the trade in Asian traditional medicines.

- The great white shark, tiger of the sea but no match for human hunters.

All these creatures are said to be vulnerable because they bear few young or are found in only a few restricted habitats, or both. But even some species and populations that reproduce in prodigious numbers or roam widely or both, like the bluefin tuna, the North Atlantic swordfish, the Atlantic cod and the haddock, are also being added to the red list, and this is the big source of controversy.

Conservationists who believe that the listing of these prolific fish is warranted point to the extinct passenger pigeon and the American bison, once nearly extinct, as examples of terrestrial species formerly thought to be so abundant as to be invulnerable. They assert that marine species as numerous as those pigeons and bison were are no less vulnerable to today's ever-more-efficient fishing methods. Fish are indeed the last wild creatures to be hunted by people on a large scale, and some conservationists say the world may be in the early stages of a marine "last buffalo hunt."

Many fisheries experts, on the other hand, scoff at such fears in the case of the wide-ranging, super-fertile bony fishes like tunas. "That's bull—that's crazy," Dr. John A. Musick, a vertebrate ecologist at the Virginia Institute of Marine Science, said of efforts to declare a number of tuna species and populations endangered. "They're severely overfished, and they should be protected," he said, but "an animal that lays millions and millions of eggs is not as likely to go extinct" as species with low reproductive rates and restricted habitats.

Dr. Musick, who agrees that some marine species other than bony fishes might be vulnerable to extinction, and Dr. Huntsman were among 31 experts from eight countries who convened at the Zoological Society of London in May to draw up the roster of fishes for the red list. Because of objections like Dr. Musick's, the red-list entries for tunas, swordfish, cod and haddock carry a notation that the listing criteria may not be appropriate for these kinds of fish. They are being included nevertheless to head off the assumption that all is well and that the populations are necessarily being properly managed. They are to be removed from the list if their numbers recover.

The red-list exercise is not intended to produce "an authoritative list of species that are about to go extinct" but rather to raise a "warning flag" for species "exhibiting symptoms of endangerment," said Dr. Georgina Mace, a conservation biologist at the Zoological Society of London, who convened the May meeting. "It's like your patient has a high temperature," she said. "It could come from just a run around the block or it could be a fever. What the scientific conservation community needs to do now is to look into those cases and, if there is a problem, work out how to manage it."

Because it is extremely difficult to study oceanic creatures, the status of only a small percentage has been assessed in even a preliminary fashion. "We just haven't looked in very many places," said Dr. Sylvia A. Earle, a marine biologist and former chief scientist of the National Oceanic and Atmospheric Administration. The species identified so far as candidates for endangerment, she said, "represent the distribution of the lookers, not the distribution of the phenomenon." Similarly, only a small number of marine species have so far been found to have become extinct in the modern era, although some biologists say that more may have done so without the extinction being detected.

In short, the investigation of marine endangerment is just beginning. Despite the division of opinion among scientists—which surfaced sharply at the London meeting, Dr. Mace said—and despite the lack of information, the question has now been placed firmly on the agenda of conservation biologists for the first time.

In the United States, the issue has taken on new urgency because of the collapse of the cod and haddock fisheries in New England, said Rolland A. Schmitten, the director of the National Marine Fisheries Service, which is charged with the overall supervision of marine fisheries in United States waters. While putting scores of marine species on the red list will be "only a first step," he said, "we welcome the message" sent by the impending action.

Another development that has brought the question to the fore is the realization that the ocean, in Dr. Huntsman's words, "isn't a homogeneous slug of water out there; it's a whole set of subhabitats, some of which aren't super-abundant." Overfish or destroy those habitats, some biologists say, and the species that live there can quickly become endangered, just as happens on land.

Dr. Michael L. Smith, an ichthyologist with the Center for Marine Conservation, a research and advocacy group based in Washington, studied the ranges of more than 500 marine species in the Caribbean and found that about 20 percent of them had been found only within a specific part of the Caribbean and often around only a single island.

If the life history of a species with a small range also makes it vulnerable to endangerment, that species is doubly threatened. Such is the case for the family of large, bass-like fish called groupers, which live on coral reefs around the world. These territorial fish never leave their immediate habitat, and it is easy for both commercial and recreational fishermen to zero in on and catch them. Dr. Huntsman, a longtime expert on groupers, said they were there for the taking for "anyone with an 18-foot boat and electronics you can buy at Kmart."

A quirk in groupers' life history, he said, makes them especially vulnerable: after several years as breeding females, the fish undergo a sex change and become breeding males. Kill the older, larger fish, as is commonly done, and the breeding males can be wiped out.

That combination of factors is landing 19 species of groupers and closely related fishes on the red list. Two, the giant jewfish and the Warsaw grouper, are termed critically endangered. Another, the Nassau grouper, is designated as endangered. The rest are classed as vulnerable.

Generally speaking, say officials of the conservation union, critically endangered means an organism faces an "extremely high risk" of extinction in the wild in the immediate future. Immediate might mean 10 years for some species, longer for longer-lived ones. Endangered means a species faces a "very high risk" of extinction in the near future—in 20 years or more, for instance. Vulnerable means a species faces a high risk of extinction in the medium-term future, which could be 100 years or so depending on the species. These categories are not comparable with those established by the United States' Endangered Species Act, which assigns imperiled species the status of "endangered" or the less serious "threatened."

Another group of imperiled fish are sea horses and related creatures like pipefish, of which 36 species are being put on the red list, most in the "vulnerable" category. Sea horses seldom move very far. They pair off for life (the male carries the young during gestation) and, like mammals, have only a few offspring, which they nurture a long time. Dried sea horses are highly prized in some Asian countries as raw material for traditional medicine, and a booming international trade has grown up to supply the skyrocketing demand, according to a recent study by Dr. Amanda Vincent, a zoologist at the University of Oxford in England. Millions of sea horses, she found, are being removed from the ocean each year, with a serious impact on their populations.

Sharks are a third major category to be represented on the red list. They, too, have few young, and fishing pressure on them has intensified as stocks of other species have been depleted. The sharks have already undergone a more intensive assessment of the kind advocated by Dr. Mace, and, as a result, several have been assigned to lower categories of risk. One, however, the Ganges shark, will still be listed as critically endangered, while the great white and the basking shark are to be listed as vulnerable.

Fish are the focus of the red list exercise, but some marine biologists are concentrating on invertebrates. Dr. Kerry Clark, a marine biologist at the Florida Institute of Technology in Melbourne, has found that because of coastal development, the number of species of sea slugs in the lagoons of coastal Florida has been reduced to 10 from about 100. Some of them appear to exist nowhere else. At least one, a small slug of a type called the sea hare, appears to be extinct, Dr. Clark said.

In California, marine ecologists have found that a deep-water mollusk called the white abalone has been overfished to such an extent that a recent sampling of 7.5 acres of abalone habitat near the Channel Islands found only three surviving individuals. These long-lived mollusks rely on the movement of water to carry sperm during reproduction. Scientists believe that the last major replenishment of the population occurred nearly 30 years ago. The reason they say, may be that the remaining abalones are too far apart for fertilization to take place. They say it is likely that the species will become extinct without human intervention.

In the case of overfishing, it has long been thought that commercial extinction, the point at which fish stocks are too small to support commercial fishing, would precede biological extinction in time to allow a species to recover if fishing is stopped. But some scientists at the London meeting argued that this was not necessarily so: some fish of high value—like the bluefin tuna, halibut, aquarium fishes, snappers, groupers and sea horses— might be pursued even though their numbers are severely reduced. Moreover, many scientists say, many less numerous species taken accidentally in fishing nets and then killed and discarded may run a risk that is less apparent but all too real.

Whatever the risk to oceanic life turns out to be upon more investigation, many experts say it is likely to increase as fishing pressure grows. The United Nations Food and Agriculture Organization estimates that 15 years from now, about 50 percent more fish than can be supplied will be required to meet the global demand for seafood.

In view of that, some scientists and conservationists agree with Dr. Carl Safina, a marine biologist with the National Audubon Society, another of the London conferees, who said, "Let's stop arguing over extinction and fix the severe depletions we have."

—WILLIAM K. STEVENS, September 1996

Throwing Back Undersize Fish
Is Said to Encourage Smaller Fry

Patricia J. Wynne

WHAT WOULD HAPPEN TO THE HUMAN RACE if tall people were hunted down and killed en masse while short people lived and reproduced freely?

"Eventually I could play in the National Basketball Association," said Dr. James A. Bohnsack, a biologist with the National Marine Fisheries Service whose dream of dunking professionally was dashed by his 5-feet-6-inch stature. "What's now considered short would soon be tall."

That is exactly what is happening to fish, Dr. Bohnsack and other biologists say. Amateur anglers have long released puny fish and kept their bigger cousins for display over the mantle. But now, these biologists say, regulations are requiring commercial fishing operators to do the same thing on a grand scale. So the fittest fish—that is, the fish most likely to pass on its genes—now is the small fish.

This theory has led Dr. Bohnsack to propose protecting big fish and their genes by placing thousands of square miles of coastal waters off limits to fishing altogether, a radical revision of the country's marine conservation programs. The proposed reserves, encompassing perhaps 20 percent of all coastal waters, would serve as genetic refuges and seed banks for restocking nearby areas.

Many biologists dismiss the idea that overfishing has caused rapid genetic changes. Nevertheless, Dr. Bohnsack's proposal has gained wide support from marine biologists. And there is a consensus that current marine conservation regulations, which rely on species' size, limited seasons and restrictions on fishing gear, have failed. The list of depopulated fish reads like a dictionary of coastal marine life. Some species like Atlantic herring and haddock show little evidence of recovery despite 15 years of conservation efforts.

"If we leave things the way they are, we're in trouble," said Dr. Greg Waugh, deputy executive director of the South Atlantic Fisheries Management Council, which regulates federal waters from Florida to North Carolina. "Just going with size limits and bag limits isn't sufficient."

In an accelerating spiral, the failure of one species often leads to the collapse of others. California's abalone industry, developed in the 19th century, relied after World War II on pink abalone. But as stocks were depleted, fishermen switched to red abalone, then to green and finally to black. A few years ago the entire industry failed. So the fishermen turned to sea urchins. Now that catch is threatened.

Coastal fisheries, the researchers argue, probably survived years of increasingly intense harvesting because they were maintained by natural refuges—areas too deep and too remote to harvest. But these reserves were moved to the front lines by armies of fishermen equipped with sophisticated gear, including satellite navigation systems, sonar, refrigeration and powerful engines.

The catch has plummeted all along the country's coasts. Fish thrown back as Lilliputians just a few years ago now are considered Brobdingnagians of the deep. Researchers at the Southeast Fisheries Science Center in Beaufort, North Carolina, a division of the National Oceanic and Atmospheric Administration, say large Warsaw grouper and other marine giants disappeared with frightening speed after a party boat fleet began plying the waters off the Carolina coasts in 1972.

"It used to be we'd hook one almost every trip," Dr. Gene Huntsman, leader of the laboratory's reef fish and coastal pelagic team, recalled in describing the 300-plus pounders that once dominated the area's waters. "We haven't seen one in years."

Between 1972 and 1985, he noted, the weight of most of the 275 species tracked by the team plunged about 75 percent on average: red porgies went from 2.6 pounds to 1.3 pounds, red snapper from 18.0 to 4.4, snowy grouper from 17.6 to 4.4, speckled hind from 19.1 to 6.6, scamp from 10.1 to 3.3 and gag from 18.0 to 4.4. Dr. Huntsman noted that the trend, which has continued, was similar if less dramatic for the commercial catch.

The commercial catch of red snapper in waters along the south Florida coast, once considered among the richest in the world, has also plummeted, dropping from 253,000 pounds in 1983 to 8,177 pounds in 1989, according to statistics compiled by the South Atlantic Fisheries Management Council, a federal agency that regulates ocean waters from North Carolina to Florida. The decline for the entire Gulf of Mexico was only slightly less precipitous, falling from 14 million pounds in 1965 to 4.1 million pounds in 1985.

While many biologists note that evolutionary changes usually take hundreds of years, evidence is growing that fishing pressure compresses generations and can accelerate the rate of such change. One study concluded that Atlantic salmon spawned earlier after fewer than two decades of heavy fishing; another that the average size of chinook salmon declined by more than 50 percent in 60 years and that the average age of maturity dropped by about two years. "Nets stop the big ones while the little ones get through," Dr. Bohnsack said. "The result is a race of miniatures."

More recently, a New Zealand study of three orange roughy spawning sites in the Pacific found that heavy fishing had a significant genetic impact.

More alarmingly, the study, published in the journal *Fisheries Research* last year, also concluded that the genetic impact could occur before the species was threatened by extinction.

But Dr. John Gold, a professor of genetics at Texas A&M University, said it was almost impossible to demonstrate empirically that excessive fishing caused genetic changes like a diminution in size. While the simplicity and logic of the theory makes it compelling, he noted that the supposed changes involved several genes whose individual effect was small, cumulative and independent.

"Bohnsack's head and heart are in the right place," said Dr. Gold, whose specialty is molecular genetics. "But the idea, although a good one, is very difficult to test."

Proponents acknowledge that marine reserves may at first increase pressure on stocks in the remaining open waters and economic hardship for nearby coastal communities. But the biologists note that tourism and scuba diving can help mitigate the loss of revenue and that, unlike bag and size limits, refuges are easy to police. The reserves' negative impact will be more than offset within a few years, they argue.

That, at least, is what happened in the Philippines, which in 1974 closed a 820-yard sliver of reef off Sumilon Island. Within two years, the mean harvest rate in surrounding waters had tripled, according to a study cited by Dr. Gary E. Davis, a marine biologist with the United States National Park Service. And after five years the sustained yield on adjacent areas had climbed to between 18.2 and 26.4 tons per square kilometer per year, among the highest reported for any coral reef in the world. Then poachers started invading the reserve. Within two years yields plummeted by 50 percent over the entire area, according to the study, published two years ago by the California Cooperative of Oceanic Fisheries Investigations, a consortium of marine institutions.

The underlying reason for the success of the reserves, which have been set up in half a dozen countries around the world, is that most reef fish rarely venture far from their home bases. And while juvenile fish channel their energy into growing, adults are disproportionately fecund.

"Unlike humans, fish get bigger and better as they get older," Dr. Bohnsack said. "It's exponential." The gonads of a single 23.8-inch female red snapper, he noted, weigh as much as those of 212 females of 16.4 inches and produce as many eggs, about 9.3 million. "A few older organisms may be more important to total reproductive output than many younger ones," he noted in a recent paper delivered at a meeting of the Gulf and Caribbean Fisheries Institute.

So current conservation methods, which spare juveniles but allow the harvest of larger fish, are counterproductive, the biologists argue. And by shortening the average life span, they increase vulnerability to mismanagement and catastrophes. "What happens if you have a natural disaster that prevents reproduction for a while?" Dr. Waugh asked.

Dr. Bohnsack said he was greeted with ridicule when he first broached the idea of a network of marine preserves three years ago in a paper for the American Fisherman's Society. "When people first heard the idea, the visceral response was 'Are you out of your mind?'" he recalled. "But after thinking about it, looking at the problem, they realize it's the only thing that makes sense."

Increasingly, a consensus is forming in support of the idea among marine conservationists. Dr. Davis said researchers at a symposium on marine fisheries in San Antonio were surprised by suggestions that the idea needed more testing, even though there remains vigorous debate over reserves' preferred size and location.

"We need to take the bull by the horns and create some of those refuges," he said.

—LINDSEY GRUSON, July 1992

Long-Line Fishing Threatens Fish and Albatrosses

HIGH-SEAS DRIFT NETS that stretched for miles and swept the ocean clean a decade ago are now banned because of their destructiveness to marine life. But scientists say the nets are being supplanted by another damaging method: fishing lines up to 80 miles long, sometimes machine-baited, that deploy thousands of hooks apiece.

Thousands of vessels from many nations, including the United States, are fishing with these long lines, as they are called, and they have become the gear of choice for catching swordfish, tuna, sharks and other wide-ranging, open-ocean species—pelagic species, scientists call them—that end up as fresh steaks at the seafood market. Scientists fear the efficiency of the long lines will deplete some of these fish populations.

"If it's a pelagic species and is in a white-tablecloth restaurant, it is likely to have been caught on a long line," said Dr. Michael Laurs, director of the National Marine Fisheries Service Honolulu Laboratory in Hawaii.

The drift nets presented a 30-mile-long wall of fine mesh that trapped nearly every living thing encountering them. The long lines are more selective, biologists say, and kill far fewer dolphins, whales and other marine mammals. But the lines are devastating many kinds of sea birds that flock to the fishing boats, nab the baited hooks before they sink and are often dragged down and drowned.

Among the most imperiled of these birds is the fabled wandering albatross. The white, black-tipped wings of this avian king of the seas have a full spread of 11 feet, the largest of any bird in the world. The birds may fly more than 500 miles a day at 50 miles an hour and cover nearly 10,000 miles on a single foraging trip—only to fall victim to a long-line hook. So many are

Long-line fishing kills fewer marine animals than drift nets do, but captures more sea birds, which flock to fishing boats and seize the bait before it sinks. Here, a separate line is deployed to try to scare away birds like the Laysan albatross.

Michael Rothman

being killed in this way, scientists say, that the wanderer's population is steadily declining.

Long-lining's effects on its intended targets are also considerable. The lines cut a wide swath through fish populations, thanks to automated equipment, satellite tracking, sonar, radar, faxed weather reports at sea and other up-to-date technologies.

This efficiency, marine biologists say, has caused a severe decline in numbers of Atlantic swordfish, which in the past were mainly taken by the far less efficient method of harpooning. One result is that the fresh swordfish steaks so familiar to American diners and supermarket shoppers are coming mostly from small, immature fish. "We are eating the babies," said Dr. Steve Berkeley of Oregon State University, a marine biologist who is an expert on swordfish.

With the decline of swordfish in the Atlantic, many long-liners have shifted their operations to Hawaii, where the fleet has greatly expanded in recent years. It joins other fleets from Asian countries, New Zealand, Australia and Chile that have made the Pacific a major arena of long-lining activity. Atlantic countries like the United States, Norway, Brazil and Argentina are also involved.

"Long-liners have filled the gap left by banning drift nets," said Ronald Smolowitz, an independent expert on the design of fishing gear based in Falmouth, Massachusetts. While it is unclear how many drift-netters have switched directly to long lines, experts say there is little doubt that the long-line fleet is growing and that it includes thousands of vessels. Estimates range from 3,000 to 10,000, including many that operate in coastal waters and set shorter lines than the large open-ocean vessels.

Long-lining is an old practice, but modern technology has vastly increased its profitability and ecological impact. The basic technique is to pay out a main fishing line from the stern of a boat. Attached to the main line are many shorter branch lines terminating in hooks baited with chunks of fish or squid. When fishing for bottom species, like cod and halibut, the end of the main line is anchored to the bottom. For shallow-water pelagic species like tuna, swordfish and sharks, the main line is suspended from a series of floats.

The advent of lightweight, synthetic monofilament lines has transformed the method, enabling a single boat out for open-ocean species to immerse 1,000 to 3,000 hooks or more. It is these lines that sometimes stretch for 80 miles, although it is more typical for them to stretch 20 to 40 miles, about the same as the now-banned drift nets. Large long-line vessels that fish deep can set out tens of thousands of hooks at once and reach depths of a mile and a quarter.

State-of-the-art long-liners use satellite positioning systems and sonar to pinpoint their quarry. Some are replacing conventional winches and hand baiting of hooks with automated systems that cut up the bait, impale chunks on hooks, deploy the line, haul it in, remove the catch, clean the hooks, neatly store the line and begin the process over. Radio beacons, strobe lights and radar deflectors mark the position of deployed gear. The technological revolution even extends to new hook shapes that snag and hold fish more securely.

As with other modern commercial fishing methods like factory trawling, this efficiency is not all to the good, as the decline of many fish stocks around the world testifies. One example in the case of long lines is the Atlantic swordfish. Scientists say that the stock is now only 58 percent of what is required to maintain a viable population, that it continues to decline and that the population of reproducing adults is only 2 to 3 percent of its unfished size. Long-liners catch lots of swordfish too small to be marketed, said Dr. Berkeley, and most of these die on the line and are discarded.

In much of the Pacific, officials say, fish stocks pursued by long-liners appear to be holding up so far, though some, like the southern bluefin tuna, are under substantial pressure. One, the bigeye tuna, has been declared endangered by the World Conservation Union, as has the Atlantic swordfish.

While long lines are less like a vacuum cleaner than were the drift nets, they still catch lots of species besides those fishermen set out for. Those of commercial value are typically kept, but many others are discarded, often dead.

Whatever the long-liners' contribution to global pressures on commercial fish stocks, it is their effect on other creatures that has stimulated the most concern. They entrap only a fraction of the number of marine mammals that drift nets did, said Sharon Young, a marine mammal specialist with the Humane Society of the United States and a member of a government study group that has recently investigated the situation in the

Atlantic. There are two exceptions, she said: pilot whales and common dolphins, one of about 50 species of dolphin, both of which are being hooked in worrisome numbers. Some endangered sea turtles are also being hooked, but it is unclear how many are dying.

It is the more readily observed deaths of albatrosses, petrels, shearwaters and other surface-feeding sea birds that most alarm conservationists. The problem mainly exists in the Southern Hemisphere and the Pacific. Government researchers in Hawaii estimate that about 3,000 Laysan albatrosses and about 4,000 black-footed albatrosses were killed when hooked by tuna and swordfish long-liners in the central North Pacific in 1994–1995.

Based on observations aboard Japanese long-liners in the southern Pacific, Australian scientists have calculated that more than 40,000 albatrosses are killed there each year. Of special concern is the wandering albatross. It is considered especially vulnerable because it is most aggressive in competing for baited hooks and because the loss of even one bird is damaging. Wandering albatrosses do not breed before the age of 10, and then produce no more than one chick every other year, which is dependent for the next year on its parents; if one dies, the other cannot provide sufficient food and the chick starves. The wandering albatross has no significant natural enemies, and this has enabled it to survive with such a low reproduction rate.

Australian investigators have found that the population of wandering albatrosses in the South Georgia and Crozet Islands, which represent 40 percent of all wanderers in the world, is declining by 1 to 2 percent a year, and mortality at the hands of long-lining is being blamed.

On balance, "drift-netting is probably worse" in overall ecological impact, said Dr. Charles F. Wurster, an ornithologist at the State University of New York at Stony Brook who has campaigned to bring the plight of sea birds to notice. But because some victims of the hooks are comparatively rare, long lines are actually a more serious threat to biological diversity, contends Sandy Bartle, a biologist at the Museum of New Zealand in Wellington, who is his government's chief expert on the sea-bird problem.

A number of remedies for the problem have been proposed, including setting out lines at night, when few birds are feeding; trailing long streamers from a line attached to a pole on the boat's stern to frighten away the birds; adding weights to make the gear sink faster; setting and retrieving gear more rapidly; and refraining from the disposal of offal while fishing so that fewer birds are attracted.

The World Conservation Union adopted a resolution last month urging the adoption of such measures. And in the Pacific Northwest, where long-liners fish for halibut, sablefish and cod off Alaska, the Seattle-based North Pacific Longline Association, a fishermen's group, has proposed making the measures mandatory under federal rules. It is in the fishermen's interest to adopt the methods, say many experts, because a hook that catches a bird will catch no fish.

Mr. Bartle says it is hard to know how widely the measures have been adopted. He believes they "have made a dent" in the problem, but says "it's not a big enough dent." It is especially not big enough for the wandering albatross, he said, since any mortality at all will insure a further population decline.

A difficulty with one of the most favored methods, the use of streamers to frighten away birds, is that the birds get used to them. In the end, some other technological solution may be required. The Norwegian fishing industry has developed a method for deploying long lines underwater so that birds cannot reach the hooks, but some experts say it has not yet proved workable in rough seas.

Even if a technological fix is found, it will mean just that many more fish caught, adding incrementally to the long lines' pressure on fish stocks. The pressure promises to increase as developing countries move into long-lining and as fishing nations around the world, supported by government subsidies, build more and more long-line vessels, Mr. Bartle said.

As long as that happens, he said, "there are going to be real big problems."

—WILLIAM K. STEVENS, November 1996

A Fish that Hails from the
Age of Dinosaurs Faces Extinction

UNLESS DECISIVE STEPS are swiftly taken to curb human predation, the coelacanth, a very rare fish once thought to have accompanied the dinosaurs into extinction, will truly die out, a team of German zoologists reports.

This gloomy assessment by Dr. Hans Fricke and his colleagues at the Max Planck Institute for Behavioral Physiology in Seewiesen follows their latest annual census of coelacanths living off the coast of Grande Comore, the largest of the islands making up the Federal Islamic Republic of Comoros. Along a five-mile stretch of coast where the fish concentrate— about one tenth of the island's coastline—the coelacanth population remained steady from 1989 to 1991, Dr. Fricke reported recently in the journal *Nature*. But from 1991 through 1994, the average number of coelacanths living in their deep Indian Ocean lava caves fell from 20.5 fish per cave to 6.5.

It is possible, Dr. Fricke says, that this alarming decline is the result of a natural population fluctuation or an emigration of coelacanths away from the survey area, but it seems much likelier that human predation is responsible.

Every year since 1989 Dr. Fricke's group has descended aboard submersibles into the coelacanths' habitat. Through thick viewing ports, the team photographs, studies and counts the coelacanths in their submarine caves, which, although close to shore, lie at a depth of about 650 feet. Coelacanths are nocturnal animals, hunting at night for bottom-dwelling prey at depths up to 2,300 feet and resting during the day in their caves. Since individual coelacanths have distinctive markings, scientists can identify and track them them year after year.

Although a handful of the big fish have been found in waters off the coast of South Africa and elsewhere in the Indian Ocean, the only known community of substantial size lives along the Grande Comore coast. Dr. Fricke believes there are about 200 coelacanths in this area, barely enough to stave off extinction.

The main problem threatening the survival of the coelacanth (*Latimeria chalumnae*) is that it lives in a coastal area heavily fished by Comorians for other species of fish used as food and sources of oil. Although the five-foot coelacanths have little or no commercial value, they occasionally grab the hooks intended for other fish and are hauled to the surface. Under a new Comoros conservation law, the landing of coelacanths is forbidden, so local fishermen usually kill them and throw them away after retrieving valuable fishhooks from their mouths.

Both European government agencies and the Comorian government have tried to halt the accidental landing of coelacanths. One measure was the mooring of "fish attractors," long, brightly colored plastic streamers, from buoys anchored farther from shore than the coelacanth habitats. The attractors only lure ordinary fish and apparently do not appeal to coelacanths.

But the attractors were anchored so far from shore that fishermen found it inconvenient to paddle their canoes to the designated area. To rectify this, international agencies helped local fishermen buy outboard motors for their canoes. But by last December, Dr. Fricke said, most of the motors had broken down and the fishermen were again working the coelacanth zone.

In their *Nature* paper, the scientists from the Max Planck Institute group propose a new approach. They suggest replacing the fish attractors, mooring them close to shore but at a depth well above the coelacanth habitat. They also suggest installing a submarine television camera in front of one of the coelacanth caves with continuous live video displayed at an information center to be built for fishermen in one of the local villages.

"Frankly, I doubt at this point that the beast can survive, but we have to try something," Dr. Fricke said.

Coelacanths are members of a very ancient suborder of fishes called crossopterygians, or "fringe-finned" fish. Although fossil crossopterygian fishes are common in sedimentary rock dated between 350 million and 60 million years old, they were long thought to have died out shortly after the end of the Mesozoic era, the age of dinosaurs. But in 1938, paleontologists were stunned to learn that fishermen off the coast of South Africa had landed a coelacanth, which captured world attention as a "living fossil."

Since then about 200 coelacanths have been caught, but none has survived capture for more than a few hours. The fish are almost always injured by fishhooks, and since they have strong jaws armed with dagger-sharp teeth, fishermen generally club them to avoid dangerous bites. It is also believed that hauling the fish up from great depths may cause injurious pressure changes in their bodies.

Until 1994 when the Comoros government banned the export of coelacanth specimens or tissue, scientists around the world were supplied with enough carcass material to study these fish extensively. Some scientists believe that modern coelacanths, which have fins below their bodies somewhat resembling legs in shape and movement, are close relatives of the line of fishes that gave rise to the first vertebrates to walk on land. This would mean that the ancestors of modern coelacanths were also the ancestors of human beings.

Another group of scientists believes, however, that land tetrapods (four-legged creatures) descended from a very different line, the lungfish.

Whatever the case, naturalists, biologists and paleontologists agree that every effort should be made to save the only species of crossopterygian known to have survived to the present day, the coelacanth.

"Everybody feels sorry about extinctions," Dr. Fricke said, "but they have become so common a lot of us just don't pay much attention anymore. The coelacanth is something special, however. It is a remarkable fish, a window into the distant past and a treasure of nature. If we let him die out it will be a tragedy."

—MALCOLM W. BROWNE, April 1995

Despite Gaps, Data Leave Little Doubt that Fish Are in Peril

REPORTS OF THREATS to the global environment often set off waves of skepticism, much of it from scientists who argue that knowledge about complex phenomena is far too limited to justify jumping to apocalyptic conclusions. The sky, they insist, is not falling.

By contrast, few have dismissed the latest reports about the dire state of the world's fisheries, despite what experts say are the extraordinary difficulties of gathering data.

Scientists, industry experts and government officials agreed at a recent United Nations conference that the evidence undeniably showed that overfishing and the destruction of habitat have caused alarming drops in marine populations; the problem is global, with implications for future food production and the economic stability of the countries dependent on fishing, and on the whole, the situation is getting worse, not better.

"I don't think there is any doubt that there are very, very severe problems out there," said Mary Harwood, a biologist and fisheries policy expert with the Australian delegation to the conference. "There is already a great deal of information about the major stocks that are in trouble, and it's quite sufficient to tell you that unless some form of strict management action is taken, these stocks will be driven further and further down toward levels from which they will not be able to recover."

The most dramatic depletions have been in the Atlantic, where commercially viable quantities of cod have all but vanished from the fabled Grand Banks, triggering the layoffs of more than 20,000 people in the Atlantic fishing communities of Canada. Off Russia's Pacific coast, the collapse of pollock stocks has brought Russia and the governments of several nations that compete for fish there to the brink of a confrontation over Russia's demand for a three-year moratorium on fishing.

Virtually every other fishing region in the world is also in peril, according to figures compiled by the United Nations Food and Agriculture Organization, the leading international agency concerned with fishing.

In some cases, the agency says, some heavily fished species are approaching not only commercial, but biological extinction. The data show that nine of the world's 17 major fisheries are in a serious decline, while four are classified as commercially depleted. The others are characterized as either "fully exploited" or "overexploited."

From 1970 to 1989, the agency says, there were significant drops in catches of Pacific perch, Atlantic redfish, yellow croaker, Atka mackerel, Atlantic mackerel and Atlantic herring. In the same period, aggregate hauls of four other important species—Atlantic cod, haddock, Cape hake and silver hake—plunged from 5 million metric tons in 1970 to 2.6 million metric tons in 1989.

The evidence, much of it visible each day in the nets of fishermen around the world, has become so obvious that it is hard to refute, Ms. Harwood and other experts say. But it is only recently that marine scientists developed the analytical tools necessary to conclude that fishing stocks were in serious trouble.

The beginning of systematic fisheries research dates to 1902, when the International Council for Exploration of the Sea, a prestigious advisory group that studies North Atlantic species, opened its doors. But the traditional approaches used in the past did not include the consideration of a vast range of environmental factors that are now routinely weighed.

The studies now include the effects of a host of human factors: the growth of large government-subsidized fishing fleets, more efficient fish-catching technology, fishing practices that kill great numbers of immature fish and "non-target" species, the destruction of coastal spawning grounds, and the large number of vessels that evade compliance with regional fishing agreements by flying the flags of nations that do not recognize international fishing regulations.

Equally important in evaluating fish populations, scientists say, are the effects of natural occurrences, like changes in water temperature and salinity, which can wipe out certain stocks, especially the fish known as small shoaling pelagics, like sardines, anchovies, pilchard and capelin.

"One thing about the overall world catch is that it fluctuates, and one reason for that is the unpredictability, to a certain extent, of small pelagic resources," said John Fitzpatrick, director of the Food and Agriculture Organization's fishery industries division. "A good example is El Niño, which has severely affected the anchovy fishery off the Peruvian coast. At times like that, if you overfish, you can endanger the whole stock."

Sudden, major disappearances of such links in the food chain can also have a direct bearing on the size and numbers of the larger fish that prey on them. For example, Canadian fishermen say that overfishing of capelin has contributed to the devastation of Atlantic cod stocks in their waters.

"Until recent times, the classical approach to fisheries research was that it was enough trouble just getting the information and working on the population dynamics of the target stock, let alone relating it to interrelationships with other species and oceanographic changes," Ms. Harwood said. "Since then, we've learned that there's more and more information that you need to make an intelligent decision about what the natural resilience of the stock is, and what a safe level of harvesting might be."

But the difficulties of stock assessment are myriad, experts say, beginning with the simple fact that fish in their natural element will not stand still long enough to be counted. There are also, many note dryly, a vast number of fish still scattered over 70 percent of the earth's surface.

The Food and Agriculture Organization conducts yearly surveys on 995 commercial marine species—everything from tuna and swordfish to squid and shrimp—from 227 spots, ranging from coastal nations to tiny administrative or political entities like the Norfolk and Christmas Islands.

Although some of the data are gathered by the agency's own regional fisheries groups, much of the information on catch sizes comes from fishing nations themselves, and it is not always reliable. This is in part because many countries, particularly in the developing world, lack the money and staffs necessary to conduct accurate surveys of fishing activities along thousands of miles of coast, much of it in remote areas.

Moreover, some data are faked by fishermen who are trying to evade conservation measures by underreporting their catch.

Despite such obstacles, "our knowledge of the state of many stocks is building up," said Dr. Zbigniew S. Karnicki of Poland, a diplomat and fisheries technology expert who is vice director of the International Council for Exploration of the Sea. "The problem is that for individual nations, it is very costly to conduct research."

Many developed nations with large fleets of huge trawlers, like China, Russia, Japan, Peru and the United States, cut research costs by stationing scientists aboard fishing vessels.

"To actually go out and find out about a stock without fishing it would be unbelievably expensive," Ms. Harwood said. "The source of a tremendous amount of stock analysis comes from basic information like catch rates—particularly catch per unit effort, changes in the catch according to the amount of fishing, and so on."

Fisheries experts say the most frustrating situations develop when fishing nations conduct huge operations and fail to collect and share data. A good example was the routine deployment of hundreds of miles of drift nets over squid stocks in the Pacific, a practice since brought under control by regulations sponsored by the United Nations.

"The netting operations exploded over three or four years, quite a sufficient time to have a dramatic impact on other stocks," Ms. Harwood said. "But there was no information at all gathered by the fleets or shared with other countries."

The United Nations fisheries conference is the first effort in more than a decade to negotiate comprehensive regulations. Such meetings are part of an effort by many governments to establish the principle that, in an age of dwindling marine resources, the once unlimited right to harvest the seas carries new responsibilities.

"In a sense it's very high-minded, moralistic stuff, but it's really what's underpinning these talks," Ms. Harwood said. "If you wish to fish, then the exercise of that right must be completely matched with a duty to know what impact you're having, to contribute to the group analysis process of that stock, and to assist in creating one if it doesn't yet exist."

—DAVID E. PITT, August 1993

Biologists Fear Sustainable Yield Is Unsustainable Idea

THE CONCEPT OF SUSTAINABILITY—the idea that renewable natural resources like fisheries and forests should be exploited with restraint so that future generations can also benefit from them—has become something of a mantra among environmentalists.

But if history is any guide, sustainability may be far more difficult to achieve than is commonly thought, three fisheries biologists argued in an article in a recent issue of the journal *Science*.

For one thing, they say, science is probably incapable of predicting safe levels of resource exploitation. And even if accurate predictions were possible, they contend, history shows that human shortsightedness and greed almost always lead to overexploitation, often to the point of collapse of the resource.

"It's quite depressing," said Dr. Donald Ludwig of the University of British Columbia in Vancouver, an author of the paper, written with Dr. Carl Walters, also at British Columbia, and Dr. Ray Hilborn of the University of Washington.

The three scientists nevertheless say there is some hope if policy makers abandon "the pretense of scientific certainty," in Dr. Walters' words, and devise commonsense management plans that protect the resource, and the economy that depends on it, against a range of uncertain possibilities.

As an illustration of the problem, they note that commercial fish populations naturally fluctuate, often unpredictably, under the influence of varying ecological conditions. Yet scientists who advise fishery managers have historically calculated a single, unvarying "maximum sustainable yield," or set number of fish that supposedly can be caught without endangering the resource.

If the number is calculated on the basis of good ecological conditions for fish, as has often happened, fish stocks will be overexploited when conditions turn bad. To complicate matters, scientists disagree on how much exploitation is safe and can never hope to reach a consensus. Catastrophe sometimes results.

The record is replete with such examples, the three experts say. In one instance, they wrote, California state scientists warned that limits should be placed on harvesting the California sardine. Fishing interests were able to produce opposing scientists who said it was impossible to overfish the species. Fishing was not reduced, the population collapsed and scientists still argue over why.

The Canadian cod fishery may be another such instance, said Dr. Walters, referring to what he called the "huge disaster" that has befallen the fishery. "That's a really good instance where the scientists were trusted to produce answers," he said. But, he added, "they were way overestimating how many fish were left in the ocean" and "the net result of putting trust in scientists was to put 20,000 people out of work" when the fishery collapsed.

Scientists may never be able to track the fluctuations well enough to advise fishery managers on safe exploitation levels. "The rate of learning might not be faster than the rate of change," Dr. Hilborn said. "That's one of the reasons that over time we might not get more certain. It's a very scary thought."

The only sensible course, he said, is to conceive of sustainability not as an effort to "hold the resource in one place" but rather as an effort to ride nature's cycles by "rolling with changes in the environment."

As an example of how this might be done, he cited the Alaskan salmon fishery. The number of spawners necessary to maintain a healthy population might range from 1 million to 15 million, depending on whether the population cycle is at a peak or in a trough. The safe course in view of this uncertainty would be to leave enough fish so that the fishery as a whole will do well whether it is a good year or a bad one. This would mean taking fewer fish than the fishery would yield in a peak year.

Quite apart from the prediction problem, the three critics contend that the human drive for wealth often leads to overexploitation in this way: Fishermen or loggers harvest as much of the resource as they can. When harvests are abundant and money is plentiful, large capital investments are made in equipment and entire local or regional economies base themselves on exploiting fish or trees. When the resource fluctuates, as in the case of fisheries, or diminishes, as in the case of forests, the vested interests plead for subsidies. The subsidies come, sometimes on a temporary basis, but the effect is to encourage further overharvesting.

Here, too, the three scientists see a possible way out. It lies in a commonsense strategy routinely practiced by financial investors in the face of uncertainty: hedging your bets and spreading the risk. When a region or a locality relies for its prosperity on harvesting a fragile and unpredictable biological resource to the exclusion of other economic pursuits, Dr. Walters said, it is as if "an investment manager told his client to put all his money into one stock."

It would be safer for both the economy and the resource, Dr. Walters said, if communities tried to create a more diversified economy. The basic strategy, he said, should be to recognize the uncertainty and risk and "build in economic flexibility and risk-spreading." Had that been done in the case of the cod fishery, he said, it would not have been the disaster it has been.

But he recognized that this strategy is not always easy to apply once a region or locality has cast its economic lot with one resource, as in the case of the bitter fight between loggers and environmentalists over the cutting of old-growth forests in the Pacific Northwest.

"Nobody's come up with a way of doing it" in the forests of western North America, he said. Tourism, for example, "doesn't really seem to cut it" as an economic alternative to logging," he said. "The number of jobs it can generate doesn't seem that large." Of efforts by the Clinton Administration to find a way out of the dilemma in the Pacific Northwest, he said, "I sure don't envy those people."

By and large, the three critics "are right on target, said Walter Reid, an official at the World Resources Institute, a Washington-based research and advocacy group, who deals with the question of sustainability. He said that was particularly true on the basic point that "to wait for a uniform consensus from scientists is a recipe for disaster."

He also agreed that sustainability could not be conceived of as some permanent, steady-state condition. Rather, he said, the normal fluctuations of nature make the sustainable use of biological resources "a moving target." But he said science was of more value in managing resources than the article might suggest, not least because although science cannot remove uncertainty, "it can certainly narrow the envelope of uncertainty."

What science does not do at all, the authors said, is take into consideration the real source of difficulty: people. Most fishery problems, Dr. Hilborn said, "are a result of not understanding how the fishermen behave rather than how the fish behave."

"The resource itself is the easier research problem," he said.

Including people and their economic needs as part of an ecosystem to be preserved is a major tenet of an emerging new American conservation ethic. Dr. Ludwig, Dr. Hilborn and Dr. Walters argue that such human attributes as the hunger for wealth and the tendency to disregard long-term consequences should also be included as part of the natural system to be studied and managed.

"Resource problems are not really environmental problems," they wrote. "They are human problems that we have created at many times and in many places, under a variety of political, social and economic systems.

These are among the guides to action they offer in summing up: Rely on scientists to recognize problems but not to remedy them. Distrust claims of sustainability, since past resource exploitation has rarely been sustainable. And confront uncertainty honestly.

"As long as you recognize your limitations," Dr. Ludwig said, "you can begin to cope with things."

—WILLIAM K. STEVENS, April 1993

The Terror of the Deep
Faces a Harsher Predator

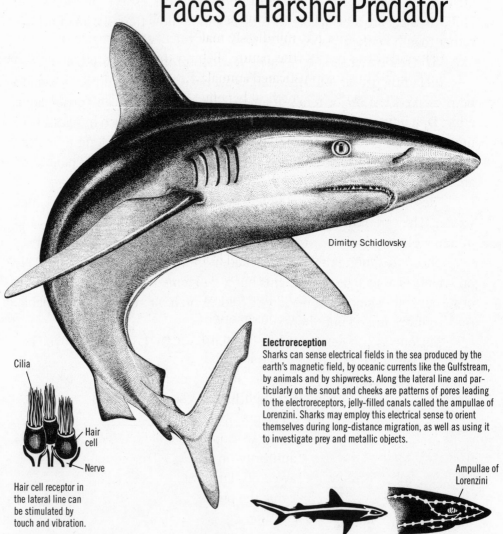

Dimitry Schidlovsky

Electroreception

Sharks can sense electrical fields in the sea produced by the earth's magnetic field, by oceanic currents like the Gulfstream, by animals and by shipwrecks. Along the lateral line and particularly on the snout and cheeks are patterns of pores leading to the electroreceptors, jelly-filled canals called the ampullae of Lorenzini. Sharks may employ this electrical sense to orient themselves during long-distance migration, as well as using it to investigate prey and metallic objects.

Cilia

Hair cell

Nerve

Hair cell receptor in the lateral line can be stimulated by touch and vibration.

Ampullae of Lorenzini

The Better to Hear You With

With two acoustic systems, sharks have a capacity for sensing underwater vibrations so acute it has been called "touch at a distance." Vibrations tens of feet away can guide them to prey or to other sharks. The sensors are in tiny openings on the top of the head, with another set along the lateral line and in pit organs.

Smell, Taste, Temperature

Mouths, nasal passages and skin surfaces, especially the snout, are lined with exquisitely sensitive receptors; not surprisingly, these hunters can sense the chemicals in blood and meat in tiny concentrations. Chemical cues may also help test the salinity of water and sense events like volcanic eruptions and chemical spills.

Eyes that Shine

Like cats, a shark's eyes shine because of the presence of a special membrane behind the retina called a tapetal plate, which reflects light back to the lens. Again like cats, some sharks also have a nictating eyelid to protect the eye.

THE SHARK, that mythic terror of the deep, has been top predator of the seas for nearly 400 million years. But sharks themselves are now being wiped out en masse by the human appetite for shark flesh, and their disappearance could disrupt the ecology of the world's oceans.

The threat comes just as scientists are reaching beyond the *Jaws* image of the shark as a primitive, mindlessly malevolent eating machine that has long shrouded the beast's true nature. Behind the legend, researchers are finding a wondrously sophisticated animal whose biology, once understood, could also yield important medical benefits.

Lured by a shark-fin and shark-meat market that has soared in the last decade, fishermen are killing sharks so fast that scientists fear entire populations are threatened. If the pace of the killing continues, the decimation of the oceans' top predator could throw marine ecosystems and food webs severely out of kilter, with possibly catastrophic consequences for other commercial fisheries.

The National Marine Fisheries Service has been struggling for more than three years to come up with a plan for stemming the slaughter. The agency is being criticized as too slow by scientists and conservationists who note that sharks' slow reproduction rate cannot keep up with the depletion of their numbers. With some conservation groups threatening to take legal action, Dr. Bill Fox, the director of the fisheries service, has set an admittedly tentative January 1 deadline for producing a regulation plan.

The great white shark—the ultimate marine predator and knife-toothed subject of movies and nightmares alike—may be one of the most threatened as a species. Great whites "are in trouble everywhere they go," which is almost everywhere in the world, said Samuel H. Gruber of the University of Miami, a leading authority on sharks. "They're out on an evolutionary limb and it's being sawed off, I'm afraid."

Shark experts recognize that the public will never come to view sharks with affection. The great white, after all, is "not an affable teddy bear whose reputation is completely undeserved," Richard Ellis and John E. McCosker wrote in a 1991 book, *Great White Shark* (Harper Collins). But biologists insist that popular myth is wrong in many respects and that there is much to appreciate about the animals.

There are more than 350 species of shark, ranging from the tiny pygmy shark to the giant whale shark. Some prey on large marine creatures; others live by straining plankton from the water.

As a group, they are seen by scientists not as dull, primitive gobblers but as extraordinary biological machines.

Their sensory system "has got to be one of the most advanced among vertebrates if not the most advanced," says Dr. Timothy C. Tricas, a shark biologist at the Florida Institute of Technology in Melbourne. They are able to hear the sounds of fish swimming at a distance. They can detect the smell of as little as one drop of fish extract in a quarter-acre lagoon six and a half feet deep. Their eyes, which equip them to distinguish colors, employ a lens seven times as powerful as a human's and include a sort of mirror behind the retina, called the tapetum, that reflects images and increases visual power.

Microscopic nerve and hair cells on sharks' skin enable them to feel the presence of moving objects in their vicinity.

A sixth sense, seated in small jelly-filled canals on the shark's snout and lower jaw, and poetically named the ampullae of Lorenzini, enables it to sense bioelectric fields radiated by other sea creatures. Through these electroreceptors a shark can detect, for instance, the heartbeat of a flatfish buried in the sand. The receptors may also allow sharks to navigate by sensing variations in the earth's magnetic field.

The information from this array of sensors, made to order for the role of top predator, is analyzed, integrated and acted upon by a relatively large and complex brain, comparable in size and function to that of supposedly more advanced animals like mammals and birds.

The immune system of sharks and their close relatives, skates and rays, appears to make them all but invulnerable to cancer and infectious diseases. Over an eight-year period, Dr. Carl Luer, a biochemist and shark biologist at the Mote Marine Laboratory in Sarasota, Florida, injected many nurse sharks with potent chemicals known to cause cancer in other laboratory animals. "We were not able to produce even early changes that would indicate a tumor was being formed," Dr. Luer said.

If scientists can find out what is responsible for this resistance, and for sharks' resistance to infection generally, it could open the way to vast medical benefits. To that end, Dr. Luer and others are investigating the immune cells of sharks to determine whether they are more efficient than those of humans and if so, why. "If we can describe the differences and similarities, we might be able to pinpoint what's responsible for sharks' success," Dr. Luer said. "But we've got a long way to go."

Some sharks, including the great white, the mako and the thresher, are warm-blooded. And contrary to the myth, sharks are not eating machines constantly on the lookout for a victim.

Dr. Gruber, working with lemon sharks in the Bahamas, has found that they feed only about every second or third day. (Lemon sharks can be studied in captivity because, unlike many other sharks, they do not have to swim constantly to force water over their gills.)

Dr. Gruber found that a shark's digestive tract works very slowly: It may take four days to digest a meal. This may be why sharks take 15 years or longer to grow to maturity, and why they grow even more slowly during the rest of a life span that can reach 100 years in some species.

"I think this has an important evolutionary meaning," says Dr. Gruber. "It allows the shark a kind of luxury that we see in many predators. It is not forced to eat all the time to keep up with demands for growth and activity. It can go on a feast-or-famine regime" as the availability of prey allows or requires.

Shark reproduction can be a somewhat violent affair, since the male typically holds the female with his teeth. Females have consequently evolved thicker skin. While some species lay eggs, in most species the fetus is nourished through a placenta, just like a human fetus, and the young are born fully developed. Not all the eggs inside the mother get fertilized, and when the unborn baby sharks develop teeth, they eat the unfertilized eggs.

In one species, the sand tiger shark, several fertilized eggs develop in each of two wombs. When the babies develop teeth, they eat each other while still inside the mother and only two survivors, one per uterus, are subsequently born.

Because sharks produce few offspring, grow slowly and mature late in life, they do not recover rapidly from overfishing and are extremely vulnerable to mass killing. This is the root of their survival plight.

Like lions, wolves and an array of other predators at the apex of terrestrial food webs, sharks face little competition from animals lower in the chain; the only thing that eats them is larger sharks. Except for humans.

The Asian market for shark fins, which soared in the 1980's, has driven the price of fins and tails to as high as $100 a pound in the case of the great white shark, the most prized species for soup ingredients. This has led to the widely deplored but equally widespread practice of "finning"—catching sharks, cutting off their fins and dumping the live but helpless bodies back into the sea.

A strong market for shark meat has also developed in the United States in recent years. Additionally, many sharks are killed and wasted when inadvertently caught by fishermen who are after tuna and other food fish.

According to the National Marine Fisheries Service, large coastal sharks, the group most affected by fishing, have been killed in numbers that exceed their reproduction rate in every year since 1987 but 1990. The peak year was 1989, when an estimated 488,000 large coastal sharks were killed. The catch has dropped off somewhat since then to 370,000 in 1991 and probably a little more this year, but killings nevertheless continue to outstrip reproduction, suggesting that the population is shrinking.

Twenty-two species, including some better-known ones, fit the category of large coastal sharks. Among them are the great white, the lemon, the nurse, the tiger, the sand tiger, the bull and three species of hammerhead. The two other categories consist of small coastal sharks and pelagic, or high-seas, sharks. The federal assessment does not consider these categories to be overfished.

A mass removal of sharks in a given locality could play havoc with ecological relationships. If an "apex" predator like the shark is removed, its loss could cascade throughout the food web.

"They affect everything below them," said Dr. Robert E. Hueter, a shark biologist who directs the federally sponsored Center for Shark Research at the Mote laboratory. "You'd have a very profound shifting" of local ecological communities.

Some years ago in Tasmania, a shark population crashed because of overfishing. As a result, octopuses on which the sharks preyed proliferated and overwhelmed their own prey, the spiny lobster. The lobster population crashed, and an important commercial fishery with it. Some scientists fear that the same thing could happen soon in Florida, where stone crabs that are the basis of a lucrative fishery are preyed on by octopuses, which in turn are eaten by sharks.

How such cascading effects would ultimately play out is uncertain. "But scientists can tell you there will be a change," said Dr. Hueter. And "whatever changes take place, you've got to learn to live with them for a long time," since it takes decades for a shark population to recover from a crash.

Conservation efforts are greatly handicapped by a lack of detailed information about the ecological effects of shark overfishing and, for that matter, about the extent of the overfishing itself. In general, said Dr. Hueter, the shark "is one of the most difficult animals on the planet to study because it's large, free-ranging and oceanic."

Dr. Fox, the director of the Marine Fisheries Service, attributes the delay in producing a shark conservation plan to "poor data on sharks and shark populations; either new data have come out or the science has caught up and caused us to take a new direction."

Dr. Fox has declined to specify what the planned shark-fishing regulations might require, but he indicated that at a minimum they would prohibit the practice of finning. The rules could also set bag limits on recreational fishing for sharks and establish quotas for commercial fishermen.

Dr. Fox said any further harm to shark populations caused by the delay in regulations "is not something they can't recover from," although the regulations might have to be more stringent to make up for lost time.

Conservationists and shark scientists say they have run out of patience.

"We won't tolerate any more inaction," said Sonja Fordham, fisheries program specialist for the Center for Marine Conservation, a Washington-based research and advocacy group. "We're prepared to take legal action to save Atlantic sharks."

Summing up the feelings of many scientists, Dr. Gruber said: "The hammering that sharks are taking simply cannot be sustained. There's so much that's fascinating about sharks that to have them killed off before we've explored the wonders of their biology and ecology, especially in ignorance and just to make a quick buck, is a crime."

—WILLIAM K. STEVENS, December 1992

Appetite for Sushi
Threatens the Giant Tuna

Pieter Arend Folkens

Unlike many other fish, tuna do not use muscles to pump water over their gills. Rather, they force a flow of water over their gills by swimming constantly. The tuna keeps its body temperature above that of the surrounding water. Scientist believe this may let it break down sugars more rapidly for bursts of energy. The high metabolic rate is aided by a high oxygen intake and a concentration of hemoglobin, which carries oxygen, as high as that of humans.

FROM MONTAUK, Long Island, as both a man and boy, Bill Camp has been sallying out into the Atlantic for a quarter of a century in one of angling's great quests: the chase for the bluefin tuna, one of the world's most magnificent vertebrates.

The bluefin is a fish to which superlatives cling. Growing to 1,500 pounds and 14 feet, it is the biggest bony fish in the world and quite possibly the strongest. It is certainly one of the fastest, as both sprinter and marathoner. By revving up its warm-blooded metabolism, it can make short dashes of 50 miles an hour. And it can cross 5,000 miles of open sea, migrating across an entire ocean, in 50 days. All in all, marine scientists say, the bluefin is a marvel of evolutionary adaptation.

In the mid-1960's, when Bill Camp was a teenager, this was the time of year when fishermen would catch as many of the biggest bluefins, called giants, as they could handle. Old photographs of the big fish, towering over the people who caught them, grace the walls of the shingled watering holes nestled around Montauk Harbor.

But those catches are gone. "There is no comparison between then and now," said Mr. Camp, the mate on captain Joe McBride's charter boat, *My Mate*. "The best guys around here are lucky to get eight or nine giants a year. We used to get seven or eight a day."

The bluefin appears to be in serious trouble, at least in the Western Atlantic. Environmentalists and some charter-boat operators say it is a victim of money lust. A giant bluefin—defined as weighing 310 pounds or more—can bring $10,000 to $15,000 on the open market in Japan, where its high-fat meat is avidly sought for sashimi and sushi. (Canned tuna is prepared not from the highly prized bluefin but from other tuna species like albacore, yellowfin, skipjack or bigeye. Yellowfin is the most common source of tuna steaks, increasingly popular in the United States.)

As the bluefin's numbers have shrunk, the price has risen, approaching $20 a pound in some years. This has prompted sports anglers to join commercial fishermen in the enthusiastic pursuit of swimming dollars.

Although international fishing quotas were imposed on bluefins a decade ago, scientists charged with monitoring the situation say there are only 10 percent as many giant bluefins in the Western Atlantic as in 1970, and the scientists say the decline is continuing. The status of the giants is crucial because only they are mature enough to spawn, and it is the spawners on which the future of the species depends.

In response, the National Audubon Society has taken the highly unusual step of formally proposing that the bluefin be listed as endangered, and that international trade in the species be banned, under an international treaty called the Convention on International Trade in Endangered Species of Wild Fauna and Flora, commonly known as CITES. It is believed to be the first time that a commercial fish has been proposed for the international endangered list.

The proposal has touched off a spirited dispute, pitting commercial tuna fishermen against the environmentalists and many sport fishermen and charter operators. The commercial fishermen challenge the scientific basis for the conclusion that giant bluefins continue to decline and assert that, in fact, aerial reconnaissance shows that they have turned the corner and that their numbers off North America are rebounding.

They maintain that while stocks of the northern bluefin tuna did decline in the Western Atlantic after 1970, the species itself is not endangered worldwide. (A separate species, the southern bluefin, inhabits the South Pacific.)

Countries adhering to the CITES treaty, which include most of the world's nations, could decide to ban international trade in bluefins. They could also allow the trade but make it subject to more stringent controls, including measures to track the trade better and hold down violations more effectively.

On a recent day when the tuna season should be at its peak, an informal survey of Montauk Harbor showed that not a single giant had been caught. The largest bluefins taken were immature fish weighing about 70 pounds. The smallest weighed about 20 pounds at most—babies, less than a year old.

These were "pathetically tiny" by bluefin standards, said Dr. Carl Safina, the director of marine conservation for the Audubon Society and a veteran tuna fisherman. Marina operators said that an average of less than one giant tuna a day was being taken last week, and that almost none had been caught in the weeks immediately before. When one is taken, it is invariably snapped up by a fish buyer, usually for export to Japan.

Trucked from Montauk to Kennedy International Airport and shipped directly to Narita Airport outside Tokyo, it goes on the auction block the next day. The payoff is handsome and immediate, and the very scarcity of the resource tends to keep prices up, making the situation advantageous for commercial fishermen.

Sport fishermen, who often sell their catches to pay their considerable fishing expenses, have transformed the nature of the sport. "The first thing a guest asks is, 'How much are they worth today?'" said Mr. Camp. "People used to come because the fish fought, not because they were worth a lot of money. It was neat then. I hate it now."

Dr. Safina tells of being out on the water and listening on the radio as one boat captain admonished another to keep one or two fish and let the rest go "so that there'll be some for tomorrow." The response came back: "Nobody left any buffalo for me."

But most times, he said, fishing is so poor, compared with years ago, that no one catches enough bluefins of any size to fill out the federally imposed sport fishing limit of four fish per person per day.

Like other tunas, the bluefin is warm-blooded, its body temperature reaching 85 degrees Fahrenheit. Its circulatory system is designed to both shed and conserve heat, as needed. Its entire makeup is geared toward speed and endurance. And in more ways than one, the bluefins are the mightiest tunas of all. Even a smallish one, says Mr. Camp, remains upright and defiant when brought alongside the boat, its characteristic blue back, golden side-stripe and silvery belly gleaming, while other tuna species roll on their sides, spent.

Spawning in the Gulf of Mexico, Western Atlantic bluefins migrate throughout the North Atlantic. Occasionally, fish caught and tagged have been found to swim from the Bahamas to Norway, from the East Coast of the United States to the Mediterranean, and from Spain to Virginia. Scientists are not sure precisely why they migrate, but believe it is because they follow changing food supplies.

The bluefin reaches sexual maturity and begins spawning at about eight years old, which is when it enters the giant class. A single female may shed 25 million eggs, only a few of which develop into fish that will survive to maturity. A bluefin's life span is more than 20 years.

For purposes of managing the northern bluefin fishery, the Western and Eastern Atlantic (including the Mediterranean Sea) are considered separate populations, since the interchange between them is relatively small. Most of the fishing in the Western Atlantic occurs off the northeastern United States and eastern Canada. In that area, the population of giant fish was found to have declined to 30,000 in 1990 from 319,000 in 1970, a decline of nearly 90 percent. By 1989, the total bluefin population had declined to 219,000 from more than 1 million in 1970.

The analysis was made by the standing committee on research and statistics of the International Commission for the Conservation of Atlantic Tunas, based in Madrid. The commission is charged with regulating the

Atlantic tuna fishery. Its scientific committee, which analyzes data from various member countries, also found in 1990 that the stock of fish under one year old was about as depleted as that of adults.

It found, too, that the stock of fish one to five years old had stabilized at about 25 percent of 1970 levels after tighter quotas on the bluefin catch went into effect in 1982, and that the population of fish six to seven years old had rebounded since 1981 to about 50 percent of the 1970 level.

The commission has established a ceiling of 2,660 metric tons, some 5,000 to 6,000 individual fish of all sizes, on the annual catch of bluefin from the Western Atlantic. The quota is divided among the United States, Canada and Japan, the three countries whose fishermen operate in the waters. The commission's scientific panel found that in 1989, the total catch was 2,800 metric tons.

The study's finding of a continuing depletion of giant bluefins has been challenged by the East Coast Tuna Association, an organization of commercial tuna fishermen and wholesalers. It says that some of the study's methods were flawed. The part of the study said to be flawed was performed by the Southeast Fisheries Center of the National Marine Fisheries Service in Miami. Dr. Brad Brown, the regional science and research director there, said that the major flaw had been corrected and the study re-run, and that there was no change in the long-range downward trend of the giant bluefins.

The tuna association asserts that its members have been catching an increasing number of "small" giants, those in the range of 310 to 400 pounds, that these fish were spawned in 1982 or 1983, and that they represent the first fruits of the commission's restrictions. They say that this might not be reflected in the commission's study because data on tuna catches, from which population trends are calculated, lag behind the present. Dr. Brown placed the lag at about a year and a half.

The commission's assessment "doesn't seem to come close" to reflecting the recovery that commercial fishermen are seeing, said Gerald Abrams, the president and founder of the East Coast Tuna Association.

Roger Hillhouse, who owns an interest in tuna fishing boats operating out of New Bedford, Massachusetts, and who flies a spotter plane looking for fish, says "small" giant bluefins have turned up in abundance off Nova Scotia, where the water is colder. These should migrate southward as the more southerly waters get colder, he said.

All parties agree that the study, which is carried out each year, is somewhat inexact and that it could be improved, and the fisheries service and the tuna association have agreed on some ways in which that might be accomplished.

Dr. Safina of the Audubon Society, who is basing his petition to place bluefins on the international endangered list on the commission's study, holds that the study is the best available estimate of what is going on. If the commercial fishermen's group "can convince the scientists that they have overlooked something and the science is revised, we will certainly accept that," said Dr. Safina.

While parties may argue about the state of the fishery, all agree that it must be preserved.

"I'd like to see it regulated while there's something left to regulate and not wait till it goes extinct," said Tommy Edwardes, dockmaster and fish buyer at the Montauk Marine Basin, where many bluefin catches come in. Mr. Edwardes dispatches many of those fish to Tokyo.

"I'm not in this for the short haul," he said. "I'd like to be in it for a long time to come."

—WILLIAM K. STEVENS, September 1991

A one-cell animal, Pfiesteria piscida, is the cause of dangerous red tides, which may be linked to farm waste. The organism has 24 different stages, including flagellated cells that feed on bits of fish, bristly cysts and several amoebic forms.

Dimitry Schidlovsky

A Spate of Red Tides
Is Menacing Coastal Seas

LIKE SOMETHING OUT OF A HORROR MOVIE, the cell from hell attacks its victims in gruesome ways, frequently changing its body form with lightning speed. The unicellular animal, called *Pfiesteria piscida,* has at least 24 guises it can assume in the course of its lifetime. It can also masquerade as a plant or lie dormant for years in the absence of suitable prey.

Armed with a voracious appetite and vast reproductive powers, the microscopic animal moves through coastal waters to kill fish and shellfish by the millions and to poison anglers and others, producing pain, narcosis, disorientation, nausea, fatigue, vomiting, memory loss, immune failure and personality changes. Its toxins are so deadly that people who merely inhale its vapors can be badly hurt.

"This thing has us scared to death," said Rick Dove, the expert who has been appointed to keep track of the Neuse River in North Carolina, part of a coastal estuary where the organism periodically goes on killing sprees. "This river is our lifeblood. If it goes belly up, everything goes belly up."

Anything but a rare organism, *Pfiesteria* has scores of toxic cousins that appear to be multiplying around the globe, mainly as algae but sometimes as zooplankton and, most conspicuously, as red and brown tides.

Some ecologists believe there is a serious global epidemic of these marine microorganisms. They fear that their toxic tides may upset the natural balance of the oceans and are urging action to reduce the runoff of sewage and other nutritive substances that seem to promote the poisonous blooms.

Other experts are more cautious, conceding that the number of reported incidents is up but withholding judgment on whether this is anything more than an upturn in a natural cycle, with more observers filing reports on the scourge.

A growing fear, intriguing but unconfirmed, is that nutrient runoff from human development, the heavy use of fertilizers and livestock farms are feeding the growth of the marauding swarms. If this is true, it bodes ill, given the global spread of the phosphorous and nitrogen from human sewage, animal waste and fertilizers that is increasingly polluting freshwater streams flowing into coastal estuaries.

"There's a correlation between increased nutrients in coastal waters and increased frequency of phytoplankton blooms," said Dr. Jane Lubchenco, a prominent ecologist at Oregon State University who studies the workings of the intertidal zone. "But that's not causation. It smells like it, but the evidence is skimpy."

Studies of the deadly blooms are accelerating, if only because the tiny killers have been found to harbor poisons one thousand times as toxic than cyanide, strong enough to kill humans.

"People in Washington are beginning to realize that you can't attack this with Band-Aids," said Dr. Donald M. Anderson, of the Woods Hole Oceanographic Institution on Cape Cod, who is an expert on red tides. "Every single coastal state has this problem in one form or another."

"Everybody agrees that the impact and number of reported events are increasing," Dr. Anderson added. "At the very least, in many areas of the world we are adding nutrients in the form of pollution into coastal waters, and that seems to be producing more harmful blooms."

Red tides are nothing new. The Bible says the Egyptians were plagued by a blood-red tide that fouled the Nile and killed fish. Homer's *Iliad* reports similar woes, and the Red Sea may have been named after the noxious blooms.

The suspicion is that natural cycles of bloom and bust are expanding into a global menace. For scientific sleuths, the challenge is to find clues that allow natural causes and cycles to be distinguished from ones that are altogether unnatural.

"The question is whether human effects are prompting an increase in the frequency, virulence or the types of organisms," said Dr. Peter Franks, a red-tide expert at the Scripps Institution of Oceanography in La Jolla, California.

Red tides, he noted, are only occasionally red, also appearing as orange, brown and even green. And they never occur literally as tides, which are the rises and falls of the sea.

The preferred terms for the phenomena are harmful algal blooms (which kill things), noxious algal blooms (which smell bad) and exceptional algal blooms (which are visually striking but do no direct harm). Unfortunately, even nontoxic blooms can kill marine animals by depleting oxygen in the water.

Moreover, the danger is entirely invisible at times. "The organisms are often in such low densities that there's nothing to see," said Dr. Franks of Scripps. "Almost every year, we have red tides that are barely visible."

A main constituent of red tides is algae, an ancient group of primitive plants dating to the first terrestrial life. The microscopic killers in most cases are algae that occur in the form of dinoflagellates, tiny unicellular organisms that usually photosynthesize and contain chlorophyll but that also have the animal-like trait of bearing twin tails, which whirl the organism forward.

Some dinoflagellates move vertically in response to light, living near surface waters during the day and diving toward the bottom to feed in nutrient-rich waters at night.

A bloom develops when the dinoflagellates photosynthesize and multiply rapidly, thriving on dissolved nutrients and sunlight. The toxins from their bodies in many cases appear to be defensive shields against zooplankton and other aquatic grazers, making the killers unappetizing.

Harm to humans often occurs when clams, mussels, oysters or scallops eat the dinoflagellates and accumulate toxins. Typically, the shellfish themselves are affected only slightly, but a single clam can sometimes pack enough poison to kill a human.

In the past decade, the number of known dinoflagellate species that are toxic has risen globally, to 55 from 22. Outbreaks once found only around the coastal areas of Europe and the United States now occur around the globe, including near Asia and South America.

Unfortunately, scientists are discovering that dinoflagellates can attack humans not only indirectly, via shellfish, but directly. No example is more gruesome than that of *Pfiesteria piscida*.

The baffling case began in the late 1980's, when scientists in North Carolina puzzled over large fish kills in which the victims were often covered with open, bleeding sores. A battery of tests ruled out fungi, bacteria, heavy metals and pesticides as the culprit. Finally, after much research, a bizarre dinoflagellate was uncovered by scientists at North Carolina State University at Raleigh and was studied in detail by Dr. JoAnn M. Burkholder, an aquatic ecologist there.

Pfiesteria piscida turned out to be not just a new species in a new genus but a whole new family of life. It kills in freshwater or seawater, but it is deadliest in polluted brackish waters where the salinity is a little less than half that of the sea.

Researchers have found that unlike most toxic algae, the dinoflagellates dispense their poisons directly into the environment, to paralyze their prey, rather than harboring them internally.

After Dr. Burkholder's co-workers got stomach cramps and other symptoms, she moved her research in April 1993 into a special room at the university meant to keep toxins isolated. But the facility leaked. Dr. Burkholder and her assistant, Dr. Howard B. Glasgow Jr., suffered a rash of illnesses, including nausea, vomiting, headaches, burning eyes, memory lapses, breathing difficulty, mood swings, impaired speech and skin lesions on their hands and forearms. During the last exposure, Dr. Glasgow was nearly overcome and ended up fleeing the facility on his hands and knees.

The lab was closed for more than a year, until July 1995, for sealing and strengthening. It was reopened only after close inspection by biological warfare experts from the Defense Department.

"Now it's a biohazard Level 3 facility," Dr. Burkholder said in an interview. "That's above rabies and on a par with AIDS. We're erring on the side of caution. We want to make sure nobody gets hurt."

Dr. Burkholder has discovered that the microscopic animal has an arsenal of disguises and lifestyles that make it extremely difficult to track through its watery environment. Of its 24 known life stages, 19 have been connected to particular survival strategies, but five remain mysterious.

"It can transform from an amoeba to a toxic zoospore in two minutes," she said. "Nobody believed that until we had people come in and watch it. Nobody had seen that before in dinoflagellates."

The animal can lay hidden on the bottom of an estuary for years in its cyst stage, awaiting an as-yet-unidentified chemical signal that says fish are nearby and prompts the dinoflagellate to come alive.

"The zoospores make toxins that are shed into the water, basically drugging the fish and making them lethargic," Dr. Burkholder said. "The toxins can rip a hole through the skin of the fish, causing bleeding sores, although some fish die so quickly that no sores develop."

In laboratory tests, Dr. Burkholder has shown that the dinoflagellate's growth and reproduction are stimulated by phosphate enrichment that is typical of the levels found in the Neuse and Pamlico Rivers in North Carolina, which have been stricken repeatedly by large fish kills and are part of the nation's second largest estuary, sheltered by the Outer Banks. Such enrichment seems to work indirectly, causing the multiplication of algae, which the dinoflagellates feed on in addition to fish.

The killer also appears to be stimulated directly by human and animal sewage, which it uses for its metabolism. Politically and economically, this is a touchy subject because North Carolina has grown in recent years to become one of the nation's top hog producers, and swine excrement is implicated in the fish kills.

"I was threatened twice last summer" by anonymous telephone callers, Dr. Burkholder said. "It was, if I knew what was good for me, I would drop the research—only more bluntly than that."

But she is continuing her research. "This is the only dinoflagellate known to have a life cycle this complex," she added. "Now that we know what to look for, I think we'll find lots of others."

Such detective work is increasingly common. After months of urgent research, scientists at the University of Miami concluded in July that a red-tide toxin caused by a different dinoflagellate was responsible for a record number of manatee deaths in Florida in 1996.

The finding was hailed as a major step toward understanding the threats facing the manatee, a 10-foot endangered mammal famous for its gentleness. Of the 304 manatee deaths during the first half of 1996, 158 have been linked to red-tide toxins, with the rest stemming from such ordinary causes as infant mortality and old age.

If the red scourge is indeed spreading and caused by nutrients from human development, experts say, then it is easy to envision a number of possible controls, like decreasing the runoffs and nutrients going into rivers. In the United States, such work has been a high priority for years, though environmentalists tend to fault the pace of remedial work by the federal government as too slow.

"Mitigation can run the gambit from better prediction, to pollution control, to improving strategies employed in agriculture," said Dr. Anderson, of Woods Hole. "If fish farmers know a red tide is coming, they can move cages."

Mr. Dove, the river keeper for the Neuse River in North Carolina, as well as a lawyer and former Marine Corps colonel, said that all the signs in his area pointed to the problem's origin lying in the rapid growth of agribusiness and factory farms and that the logical solutions centered on them as well.

"This river's been out of whack since 1991," he said. "Around that same time, we got big in the hog industry. We don't know if that's the only problem. But we know that they're producing more fecal waste than all the people in the state of New York."

"This is one of the most beautiful rivers in the country," he added, speaking of the Neuse. "Now it's a great embarrassment. It should never have been allowed to get to this point. *Pfiesteria* has been around since the beginning of time. But it's sure out of control now."

—WILLIAM J. BROAD, August 1996

6

SHELLFISH
AND
OTHERS

The oceans hold a host of creatures besides fish, many with strange forms and habits that biologists are only beginning to understand.

There are the squids, delicate-bodied masters of subterfuge and decoy. Undersea diving capsules have recently filmed them in the deep waters of their native habitat. But the giant squid, *Architeuthis*, remains to be captured alive or observed in its natural lair.

There are cone snails, which have developed a repertoire of neurotoxins that a surgeon might envy for their precise targeting of various regions of their victims' brain. The corals too are masters of toxin warfare.

After millions of years of evolutionary warfare between prey and predator, weapons and strategy have advanced to surprising heights of subtlety. The following articles offer a taste of the rich lifestyles of the ocean's creatures other than fish.

Deadly Snails Take Pinpoint Aim with Diverse Toxins

Deadly Harpoonist

Each species of cone snail has its own collection of amazingly specific toxins. One issued by Conus purpurascens *to paralyze an unwary fish that has bitten a line that looks like food. The line comes with a harpoon attached, and the fish is first rendered stiff, then reeled into the snail's maw.*

Mick Ellison

HIDING IN CORAL REEFS in the Philippines, venomous cone snails have lured collectors with their brilliant and intricately patterned shells. The snails, just inches in length, have intrigued biologists because they can catch fish as large as they are. And increasingly they are attracting neurobiologists because it turns out that the snails make thousands of toxins that lock onto crucial molecules of mammalian nervous systems with pinpoint precision.

The cone snail toxins can knock out particular molecules needed for the transmission of certain nerve impulses while leaving similar molecules alone. In addition to their use in studies of how the nervous system works, they may lead to therapeutic drugs that avoid the undesirable side effects that occur when a substance used to block one molecule inadvertently blocks similar molecules.

195

The array of the toxins' effects is like nothing ever seen before. Inject one toxin into the brains of rats and they fall asleep. Inject another and they start to scratch themselves; another and they swing their heads; yet another and they go into convulsions.

Now the first of these toxins is being tested clinically at 30 medical centers across the United States by the Neurex Corporation of Menlo Park, California. The toxin, which blocks the transmission of pain impulses up the spinal cord to the brain, is the focus of a study that will enroll 300 patients with AIDS or cancer who have intractable pain. Dr. Paul Goddard, president and chief executive officer of Neurex, said he was extremely encouraged by the results so far. He said the company expected to complete its enrollment of patients by the end of the year.

Dr. Ponzy Lu, a biochemist at the University of Pennsylvania who is interested in the structure of small molecules, said that with most of the toxins yet to be characterized, "I think there will be a lot more coming out of this."

The slowly unraveling discovery of the cone snail toxins began with a molecular biologist's desire to do science in his native Philippines. The scientist, Dr. Baldomero M. Olivera, now at the University of Utah, said he had no idea when he turned to the snails how fascinating they would be. He had received a Ph.D. at the California Institute of Technology and then done post-doctoral work at Stanford University, where he was a discoverer of ligase, an enzyme that is crucial for recombinant DNA experiments. Finishing

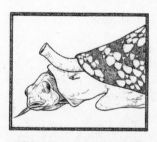

his post-doctoral work in 1969, he decided to go back to the Philippines to study molecular biology.

"I ended up at a laboratory at the University of the Philippines that had no equipment," Dr. Olivera said, so he decided to study something that was readily available. "I figured it would make a nice kind of short-term project to purify out what made these snails so deadly."

Dr. Olivera had collected shells as a child and knew about the venomous marine snails in the waters near his home. The shells of cone snails have attracted collectors for years and have killed about 30 hapless shell hunters with their poisons. Their shells have been so treasured that one two-inch-long shell brought in more money at an auction in 1796 than a painting by Vermeer.

There are about 500 species of cone snails in the world's tropical ocean waters, and about 70 of them hunt fish. Some bury themselves in the sand, waiting for a fish to come by. When a lurking snail senses a fish, it throws out a slender tube from its mouth and wiggles it. The tube can be transparent or it can be colored a brilliant red, a soft amber or a velvety purple, depending on the species of snail. But it looks, Dr. Olivera said, for all the world like a fishing line. When a hapless fish swallows the line, a poisonous barb comes jetting out, paralyzing the fish with deadly toxins. The snail reels in the fish and swallows it whole. After about an hour and a half, it spits out the bones, the scales and the barb it used to kill the fish.

Other cone snails catch sleeping fish with the biological equivalent of a fisherman's net. They open their mouths wide, so the fish drift in. Then they jab the fish with a poisonous barb to kill them.

"We started out basically looking at things that cause paralysis of fish," Dr. Olivera said. "The first surprise was that there were really a lot of things in venom that cause paralysis. If you look at sea snakes, which also eat fish, almost all of their paralytic activity is in one component of their venom."

He characterized a few of the snail toxins that paralyze fish. One is like the toxin that cobras and other poisonous snakes make. It blocks a molecule at the junction between a nerve and muscle, preventing nerve impulses from pass-

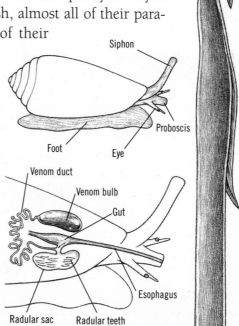

A Pump for the Snail's Poisons
The venom is made in the venom duct, a long tube, and is delivered by the harpoon-like radular tooth, shown in extreme closeup. It is pumped from the duct by the muscular venom bulb. Up to 50 such teeth are stored in the radular sac: when needed, one tooth is moved into the proboscis.

ing to the muscles. Another resembles the toxin made by the Japanese fugu, or puffer fish, which paralyzes muscles by preventing electrical signals from spreading through the muscle from the place where the nerve touches it.

Dr. Olivera said the snail's poisons hit its prey in so many places that an equivalent brew made from other known toxins would have to include curare, the poison that natives of the Amazon put on their arrows, tetrodotoxin, the poison of the puffer fish, and the botulism toxin.

He said the snails probably developed the diverse and powerful toxins because they had to paralyze their prey quickly. "Once a snail harpoons a fish, it can't have the fish jerking around," Dr. Olivera said. "That would attract other predators. The snail is very vulnerable in its feeding."

So, he said, the snails developed toxins consisting of small protein fragments, or peptides, that not only paralyze fish but shock them in much the way that an electric eel shocks its prey. "The first thing you see is that the fish's fins stiffen up," Dr. Olivera said, from the chemical equivalent of an electric shock. "It's rigid, as if it is being shocked. That gives time for the toxins that wipe out the nerve–muscle connections to spread through its body. It's a one-two punch."

But after studying the two toxins, which seemed not too different from toxins found in other species, Dr. Olivera all but abandoned his study of the snails. By the early 1980's, he had moved back to the United States to do research in molecular biology, ending up at the University of Utah. He had decided that the snail toxins were "interesting but they weren't *that* interesting."

Then, a few years later, an undergraduate student, Craig Clark, came to him with a proposal. "He got, in retrospect, a brilliant idea that instead of looking at paralysis, we should inject the peptides directly into the central nervous system," Dr. Olivera said. "To be honest, I didn't really think we'd learn anything from that, so I tried to dissuade him. But he did the experiment anyway and he made this incredible discovery that the venom was laden with components that made mice do different things." In analyzing the venoms, the researchers had separated them into components or fractions. "We just tested every fraction that had some venom component in it and almost everything scored," Dr. Olivera said.

That was the turning point, when Dr. Olivera began to consider devoting his laboratory to the full-time study of the snail venoms. "About six years ago, I kind of quit everything else," he said. "It became clear that this project was worthy of all of our time."

He learned that each of the 500 species of cone snails seemed to have its own distinct collection of toxins and that these tiny, rigid peptides are just a fraction of the size of poisons made by snakes, scorpions or spiders.

And they were amazingly specific. For example, one toxin, which he named the omega toxin, prevents certain nerve cells from releasing their signals by blocking calcium channels, tunnels in nerve cells that open to allow calcium to enter. But the toxin binds only to one subset of calcium channels, the N subtype, discovered a decade ago by Dr. Richard Tsien of Stanford University. The N channels are found only in nerve tissue. In contrast, calcium channel blockers that are used as drugs for heart disease patients block calcium channels in smooth muscle, skeletal muscle and cardiac muscle, as well as in nerves.

It is the omega toxin that the Neurex Corporation is studying to relieve intractable pain. Dr. Goddard said that about 150,000 cancer patients had severe pain that was not relieved even by morphine and that an additional 700,000 people had intractable pain from conditions like shingles, a viral infection of the nerves. His company's current study involves patients with cancer and AIDS who are receiving the omega toxin directly into their spinal cords.

But most of the snail venoms remain to be studied, Dr. Olivera said. The venom that contained the omega toxin, for example, has about 80 other components that have biological activities. "I'd say we know what about a dozen of them do," he said.

Some investigators, like Dr. Tsien, who described the snail toxins as "a very important scientific tool," are using them to understand how the nervous system is put together. Dr. Tsien is particularly interested in the N-type channels and said he could now locate those molecules by using omega toxin as a probe. In the spinal cord, the N-type channels transmit pain impulses. But the channels are also found in the brain, where they regulate the heart rate and blood pressure, Dr. Tsien said. That is why Neurex is injecting omega toxin directly into patients' spinal cords, where the toxin sticks fast to the nerves involved in pain sensations and never migrates to the brain, he said. If the toxin was injected into the brain, he said, patients would faint if they tried to stand up suddenly.

"These snails are really a kind of gift from nature," said Dr. J. Michael McIntosh, a psychiatrist at the University of Utah who is studying the toxins as possible treatments for psychiatric illnesses. "They are little chemical factories that are in essence doing what drug companies are trying to do. They've created thousands of chemical compounds and refined them to be exquisitely sensitive and potent."

—GINA KOLATA, August 1996

Australian Navy Helps Endangered Giant Clam to Relocate

THE AUSTRALIAN NAVY has stepped in to help the giant clam, an endangered species that has been driven to extinction in waters off many South Pacific islands and heavily fished on the Great Barrier Reef by poachers.

Over a recent three-day period, a heavy landing craft, the *Tarakan,* was used to transfer 3,000 clams weighing a total of more than 20 tons from experimental beds off Orpheus Island, where they had been artificially bred by marine biologists, to an undisclosed location on the outer Great Barrier Reef.

John Lucas, a professor of zoology at James Cook University at Townsville who coordinated the research on the artificial breeding of the giant clam, *Tridacna gigas,* said the operation was believed to be one of the biggest relocations of marine organisms in quantity and weight.

Eighteen divers from the *Tarakan* and the Great Barrier Reef Marine Park Authority retrieved clams at high tide from a seabed about 100 yards from the shore of Pioneer Bay, and lifted the mollusks, some measuring two feet across and weighing 40 pounds, into dinghies powered by outboards. As each dinghy was loaded with 10 to 20 clams, depending on their size, it returned to the *Tarakan,* anchored half a mile away. The bivalves were gently unloaded by hand, placed in two-foot plastic tanks filled with sea water, and then covered with shade cloth to minimize stress.

White buoys marked a rectangular area of about 200 square yards where the divers, wearing thick gloves for protection from coral, collected the clustered clams. One diver, Mike Bugler, a project officer with the park authority, said some mollusks were up to seven years old.

"The water is clear and ideal for diving, but the bigger and heavier clams can be difficult to grasp on the seabed and then bring to the surface," Mr. Bugler said.

With daytime temperatures in Pioneer Bay averaging above 90 degrees, the navy began lifting the giant clams out of the water about an hour after sunrise to avoid stress to the mollusks during the hottest part of the day, and they were in the tanks aboard the *Tarakan* within 30 minutes of being lifted from the seabed.

After three hours, with 2,000 clams resting in more than 30 tanks, the *Tarakan's* crew erected an aerial spray eight feet above the deck to deliver a fine mist of sea water to an area normally crowded with army tanks, trucks and military equipment.

Tridacna gigas, prized by poachers for its shell and the high level of protein in the meat of its adductor muscle, is listed by the International Union for Conservation of Nature as an endangered species. To protect the clams and allow them to reproduce undisturbed, the place on the Great Barrier Reef where they were moved was not disclosed, but Lieutenant Rick Watson of the navy, who commanded the *Tarakan,* said it was several hours' sailing time from Pioneer Bay.

An estimated 12,000 of the artificially bred mollusks remain in the experimental beds off Orpheus Island, 50 miles north of Townsville on the north Queensland coast.

The relocated clams will assist research on population outbreaks of the crown-of-thorns starfish, known for its destructive grazing of reefs. Recent satellite studies of reefs and currents, and other modeling of dispersal of starfish larvae, have indicated possible "source" and "sink" reefs, with currents from the source reefs tending to carry the larvae from released populations to the sink reefs.

"The clams are being used to test this by loading some source reefs with these groups of genetically related clams from the same stock," Professor Lucas said. "If the source and sink hypothesis is true, then on the sink reefs we may see increased recruitment of juvenile clams, which are shown to be genetically related to the parent stock."

James Cook University is helping the South Pacific island countries of Fiji, Tonga and Cook Islands develop their own clam farms. It is supplying one-year-old clams grown at hatcheries on Orpheus Island, and villagers can harvest the clams when they grow to maturity over three to four years.

"The first two years is the most difficult as growth is fairly slow and the islanders are going to need encouragement during those early years, but a reliable estimate of yield at about five years old is eight tons of clam meat per acre per year," Mr. Lucas said.

At four to five years old, about 32,000 giant clams can be farmed in one acre of clear, warm, offshore tidal waters, enabling Pacific Islanders to preserve their clam resources, and at the same time farm the mollusks for a food source.

The project has advanced mariculture methods significantly with the development of a simple technology for large-scale marine culture methods in five phases: spawning, hatchery, nursery, ocean nursery and growth to maturity.

The bivalves are hermaphrodite, containing both male and female sex organs, or gonads, and they usually spawn into the seawater during the summer months, with some spawning up to a billion eggs.

"It is one of the most fecund animals in the world and yet it is never naturally abundant," Mr. Lucas said. "It is a real mystery why *Tridacna gigas* spends so much energy and effort on reproduction, and then has so few progeny."

The oldest giant clam, weighing more than half a ton, was found on a reef near Townsville, and sectioning of the shell established it at 63 years old.

—NINA BICK, April 1993

Roots of the King Crab

KING CRABS MAY BE MAJESTIC in their dimensions and princely in the sums their succulent legs command at the fish market, but their origins turn out to be quite humble. Scientists have learned that the giant shellfish evolved from mere hermit crabs.

Some biologists examining the details of their anatomies had long believed they were distantly related, but a new molecular comparison of their genes shows that they are much more closely linked. Rather than being members of the same super-family, they should be counted as first cousins in the same genus. Scientists now suggest that king crabs began evolving from a hermit crab progenitor about 19.5 million years ago.

"We thought we'd find relatedness, but the degree of it was a total surprise to us," said Dr. Clifford W. Cunningham of the University of Texas at Austin, an author of the latest report, who collaborated with researchers at Yale University. The new paper appears in a recent issue of the journal *Nature*.

Hermit crabs, a diverse group of shoreline organisms, are small and defenseless and must live in other species' shucked-off shells to avoid predators. As they mature they move into progressively larger shelters, but are limited in girth by how big a shell they can scavenge.

Dr. Cunningham and his colleagues suggest that some members of the hermit crab group called Pagurus began developing a hardened skin across the bodies. That self-made armor freed them from the need to fit themselves into available housing. Over generations the well-sheathed crabs grew ever broader, their mail ever tougher and their figures ever more crab-like. Eventually, they reached their current size, about three feet wide and about 20 times the size of an adult hermit crab.

A hermit crab has an asymmetrical abdomen that allows it to squeeze into the whorls of a snail shell. The king crab, too, under its tough outer cuticle, has a belly curved to one side.

—NATALIE ANGIER, February 1992

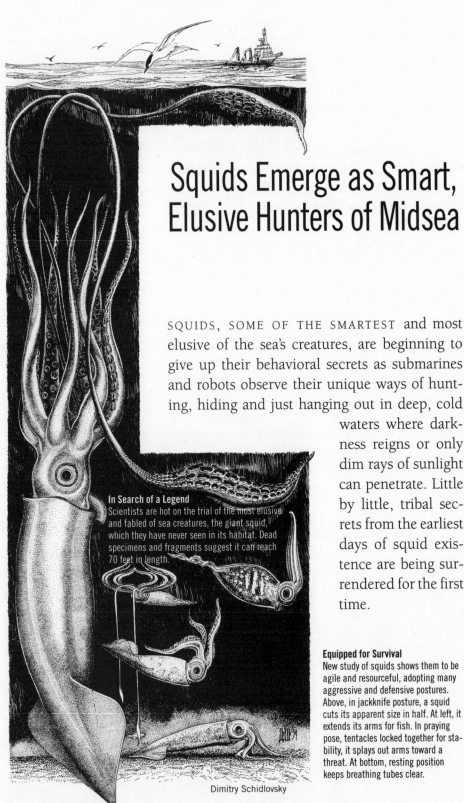

Squids Emerge as Smart, Elusive Hunters of Midsea

SQUIDS, SOME OF THE SMARTEST and most elusive of the sea's creatures, are beginning to give up their behavioral secrets as submarines and robots observe their unique ways of hunting, hiding and just hanging out in deep, cold waters where darkness reigns or only dim rays of sunlight can penetrate. Little by little, tribal secrets from the earliest days of squid existence are being surrendered for the first time.

In Search of a Legend
Scientists are hot on the trial of the most elusive and fabled of sea creatures, the giant squid, which they have never seen in its habitat. Dead specimens and fragments suggest it can reach 70 feet in length.

Equipped for Survival
New study of squids shows them to be agile and resourceful, adopting many aggressive and defensive postures. Above, in jackknife posture, a squid cuts its apparent size in half. At left, it extends its arms for fish. In praying pose, tentacles locked together for stability, it splays out arms toward a threat. At bottom, resting position keeps breathing tubes clear.

Dimitry Schidlovsky

206

Previously seen as primitive and lethargic, deep-sea squids turn out to show surprising alertness and alacrity. Most boast a range of bioluminescence and subtle coloration that can change quickly, often to elude predators and perhaps to attract mates as well.

Long tentacles, once seen mostly as grippers, also turn out to work as fishing lines and lures. A squid's arm will hold a thin tentacle, letting it run over the arm's tip and dangle far below its body. Some species then flash a light at the tentacle's end to attract prey, grabbing hold of the next meal with the tentacle's suckers.

Traditionally, squids have been thought of as creatures of the ocean's middle levels, always jetting about or floating in a state of neutral buoyancy. But it turns out that some species take breaks on the bottom, resting their arms in such a way ("on their elbows," an expert jokes) so tubes for breathing and propulsion stay clear of obstructing mud.

In some cases scientists are learning whole sequences of maneuvers meant to escape predators. Squirts of ink, it turns out, can do much more than blind a pursuer.

Even the giant squid, the largest and most legendary of the race, reaching lengths of 70 feet, is being tracked more closely than in the past. Still, scientists have yet to snare the creature, while fishermen sometimes do so by accident. A scientific expedition off the California coast has been trying to track one of the giants down, churning through dark waters with an unmanned robot, seeking to film the behemoth in its lair.

"The curtain's going up," Dr. Clyde F. E. Roper, curator of mollusks and a squid expert at the National Museum of Natural History at the Smithsonian Institution, said in an interview. "In the last 10 years we've really come a very long way in understanding the behavior of a lot of these animals. We're finding tremendous diversity."

Video images of rare, deep-water squids taken by camera-toting robots and submersibles are becoming so common that Dr. Roper and Dr. Michael Vecchione, a squid expert at the National Marine Fisheries Service of the National Oceanic and Atmospheric Administration in Washington, are putting them together in a computerized database that covers scores of species. Among other things, analysis of this image bank is revealing a host of previously hidden behaviors.

"There's much we don't know," said Dr. Vecchione. "But we're starting to reach a critical mass. Enough dives have been made so that we're starting to see some patterns. What's happening is very exciting."

Squids are cephalopods, close cousins of octopuses and cuttlefish and distant relations of clams and oysters. Their elongate bodies have rear fins. Their large eyes rival human ones in complexity, while those of the giant squid are sometimes bigger than a dinner plate.

Of their 10 arms, eight are short and meaty and two, referred to as tentacles, are usually much longer and thinner. The ends of tentacles are often expanded and covered with numerous suckers. Squids range in length from less than an inch to the 70 or so feet of the giant, and perhaps longer.

While squids that roam the sea's surface have been studied since the time of the ancient Greeks, their deep-sea relatives for the most part have remained a riddle. Hints of behavioral richness came as the first catches from the deep revealed that some squid bodies were bedecked with arrays of bioluminescent lights.

"Nothing can be even distantly compared with the hues of these light organs," wrote Carl Chun, a turn-of-the-century squid expert who was awed by a particular find. "One would think the body was adorned with a diadem of brilliant gems." The colors included ultramarine, pearl, ruby red, snow white and sky blue.

Early discoveries were limited because of the available tools, mostly nets and trawls. These missed much life. At times, they would capture some of the delicate squids of the deep, but leave them maimed or mangled. Dr. Roper said nets and lines towed through deep water undoubtedly set off great flashes of bioluminescence among jellyfish and other gelatinous creatures, warning of human onslaught.

"Any squid worth his salt is going to do a few flips and be out of the way," he said. "The only cephalopods we catch in midwater trawls are the slow, the sick or the stupid."

The advent of deep-diving robots and manned submersibles has changed all that. While such vehicles were first used to study the bottom, in the past decade or so they have increasingly examined the riot of life in the sea's middle waters. So too, the advent of small, high-quality video cameras, accompanied by special lights, have allowed scientists to readily record and analyze the squid dance.

"We've found them doing all kinds of things we thought they couldn't do," said Dr. Roper.

Pioneers in such work include Dr. Roper of the Smithsonian and scientists at such places as the Woods Hole Oceanographic Institution in Woods Hole, Massachusetts; the Harbor Branch Oceanographic Institution in Fort Pierce, Florida; and the Monterey Bay Aquarium Research Institute, in Pacific Grove, California.

Dr. Vecchione of the National Marine Fisheries Service, who is also in the field's forefront, said the new undersea perspective was rewriting the books on squid distribution and abundance in addition to behavior. "There was one species we thought was rare because we could never catch it by net," he said. "But we went down and it's everywhere. It's as common as cattle when you go down in a sub."

A recent dive conducted by the Monterey Bay Aquarium Research Institute found a wealth of odd squids in the deep waters of the Monterey Canyon, just off the California coast. Tethered to a 110-foot ship by a long electronic cord was a two-ton, unmanned robot the size of a small car that darted through the depths and continuously sent back ghostly images that were monitored in a comfortable control room.

Among the experts on board was a squid specialist, James Stein Hunt, a doctoral candidate from the University of California at Los Angeles, who works at the institute.

More than a quarter-mile down, the robot came across a delicate squid, its dark ink sack clearly visible through its transparent body. The animal, probably a juvenile and perhaps six inches long, was moving downward at a leisurely pace, its arms folded into a tight cone as its paper-thin fins undulated back and forth in a fascinating rhythm. It was performing no evasive maneuvers, despite the robot's lights. The camera slowly panned the squid's body, revealing minute details.

"It looks different," said Mr. Stein Hunt, who has been studying the region's squid for two years.

The robot's pilot, delicately working a joystick, slowly positioned a sampling jar over the diving squid and then sped up the robot's rate of descent. As the jar encircled the animal, a lid swung shut, capturing the squid for later inspection.

"It probably will turn out to be Galiteuthis," said Mr. Stein Hunt, referring to a genus name of a fairly common deep-sea squid. "It's just that, because these things are so rarely seen, especially in the younger stages, we don't have a good record of what they look like. Most of the information we have historically on younger squids was taken from trawling samples, when the animals were chewed up a bit."

Suddenly, something dark and black sped close by the robot's camera, quickly disappearing from sight. "It looked like a Vampiroteuthis," said Dr. Bruce H. Robison, the institute's chief scientist, referring to a rare kind of living fossil that is half-squid, half-octopus. The creature's name arises from its resemblance to a vampire. Its arms are covered with dark cape-like webbing, and extending from beneath the cape are two fang-like tendrils.

The robot's eye fell on a pinkish squid that started moving away rapidly, flapping its fins, trying to escape the onrushing machine. It performed an abrupt left turn. We followed. It sped up. We struggled to keep pace. It succeeded in pulling ahead.

"He's going up," said Dr. Robison. "They usually head down."

Abruptly, the vanishing squid inked and inked again, leaving tubular patches of inky material floating motionless in the water. We could see no trace of the animal itself. It had escaped the planet's most deadly predator, man.

"The ink is a pseudomorph," Dr. Robison noted. "It's supposed to look like the body of a squid" to fool a predator into attacking the ink rather than the escaping animal.

"A long thin squid makes long thin ink," said Mr. Stein Hunt. "A short round squid makes short round ink." In darkness, he added, shimmers of bioluminescence from microscopic fauna in the agitated water would surround the dark ink, highlighting the ruse.

A half-mile down, we spied a squid that went into a strange pose as the robot approached. Head down, it locked its tentacles together, as if praying, while splaying its arms outward toward the robot. Mr. Stein Hunt said the posture was defensive, readying the squid to battle an unknown threat. The locked tentacles, he said, apparently stabilized the squid and kept it from tumbling as it drifted slowly downward.

More than a half-mile down, we came upon a small squid relaxing in a vertical pose, its arms pointed down, its fins barely moving. Just visible along either side of the arms were faint translucent lines.

"You know how some aircraft have fins up by the nose?" Dr. Robison asked. "Those are canard fins. We've discovered that a number of squids have little flared, flange-like additions to their arms or tentacles that they use like canards. We think it allows them to steer better."

Later, as the ship headed back to port, Mr. Stein Hunt played a video-tape of earlier findings. One sequence showed a squid fishing, its arms held up in gentle arcs, its head down, its long tentacles dangling straight down, reaching far below its body. Another fishing scene showed a squid lying hor-izontally, one of its arms dangling a long tentacle like a lazy fisherman.

The most astonishing sequence showed Vampiroteuthis, the living fos-sil. It was the first known filming of the half-squid, half-octopus in its dark habitat. Surprisingly, given its primitive nervous system, the brownish-red creature swam with great dexterity, flapping large thin fins that looked like wings. It trailed a long filament, whose function is unknown but probably sensory.

A close-up view showed the animal perfectly still, slowly opening and closing its large arms with a sinuous rhythm, revealing them to be joined by thick webbing. Visible at the animal's front was a huge eye, eerie and blue. At the center of its web was a beady mouth, held motionless.

"Everything that was written about this animal was that it was a slow, sluggish sort of cephalopod," said Mr. Stein Hunt. "But when you actually see it alive, doing flips and carts and trailing that thing and moving all over, it's clear all that has to be rewritten. We have to rethink the animal."

Mr. Stein Hunt was also clearly eager to the learn about Architeuthis, the giant squid. He and his colleagues from the institute are now looking for the beast around the Pioneer Seamount, which rises from the seafloor some 50 miles off the California coast.

"I would do nearly anything to see one of them alive," Mr. Stein Hunt said during the Monterey dive. "That's every teuthologist's dream."

Teuthologists (pronounced tooth-AHL-o-gists), as squid experts call themselves, from the scientific Latin for squid, recently gathered near Naples, Italy, to compare notes and images. Among those attending the meeting were Dr. Roper, Dr. Vecchione and Mr. Stein Hunt, who showed his tape and amazed the audience with images of the living fossil.

"They completely blew my mind," said Dr. Vecchione. "Nobody suspected that it acted like that—buzzing around in circles and swimming rapidly. Usually they've been pictured as drifting with their arms spread out."

Other secrets of squid life were revealed in a joint paper and video presented by Dr. Vecchione and Dr. Roper. It turns out that a wide variety of related deep-sea squids adopt similar postures, in particular one known as the cockatoo pose. Here, the squid's arms are thrown out of the usual linear arrangement, jackknifing up and back so they fall close to the body. While looking humorous, like a cockatoo with its elaborate head plumage, the maneuver, the scientists speculate, lessens a squid's overall body size and the chance of being observed by a predator below.

"The orientation of surface squids is usually horizontal," said Dr. Roper of the Smithsonian. "But it turns out these midwater guys are literally hanging out in all kinds of odd positions."

Dr. Roper said he expected that eventually even the giant squid would come under scientific scrutiny and give up some of its behavioral secrets. But so far, despite numerous attempts, no teuthologist has ever come across one in its habitat, whatever that may be. No one has ever seen how giant squids move and capture their prey, or how they react to potential predators.

"They're still a mystery," he said. "People have gone out specifically looking for the giant squid and have never found them. One of these days it's going to work. One of these days somebody is going to see one alive."

—WILLIAM J. BROAD, August 1994

Scientists Close In on Hidden Lair of the Ocean's Fabled Giant Squid

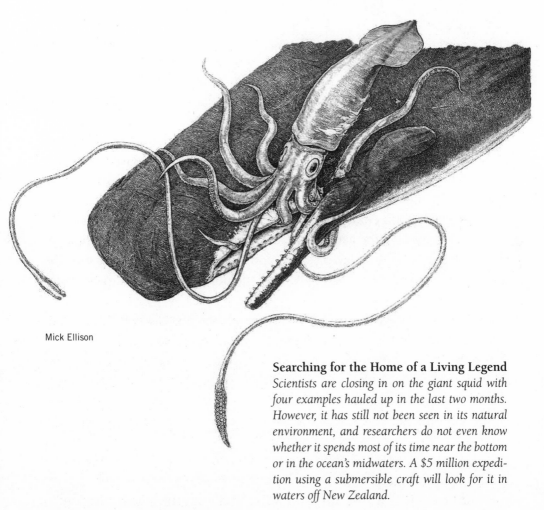

Mick Ellison

Searching for the Home of a Living Legend
Scientists are closing in on the giant squid with four examples hauled up in the last two months. However, it has still not been seen in its natural environment, and researchers do not even know whether it spends most of its time near the bottom or in the ocean's midwaters. A $5 million expedition using a submersible craft will look for it in waters off New Zealand.

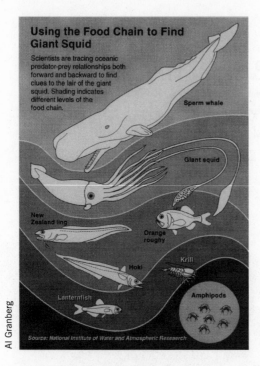

Using the Food Chain to Find Giant Squid

Scientists are tracing oceanic predator-prey relationships both forward and backward to find clues to the lair of the giant squid. Shading indicates different levels of the food chain.

Sperm whale

Giant squid

New Zealand ling

Orange roughy

Hoki

Krill

Lanternfish

Amphipods

Source: National Institute of Water and Atmospheric Reseaerch

Al Granberg

THE LAIR OF THE GIANT SQUID is a staple of novels and horror movies, and perhaps of nightmares. But for biologists it is a mystery.

No one has ever observed the beast in its natural habitat, despite decades of probing the sea's dim recesses. Fishermen towing nets through the depths have snagged giant squids on occasion, and dead or dying ones have been known to wash ashore, often half-eaten by birds and sea life. But more than a century after the giant squid and its supposed habitat were featured in *20,000 Leagues Under the Sea,* the surprising truth is that very little is known of the deep monster and how it eats and rests, courts and mates, swims and behaves.

That may soon change, however. Scientists have made much progress lately in discovering the giant's den. In the last two months alone, biologists and fishermen around New Zealand and Australia have cast nets into the deep and caught four of the big squids, including one of the largest males ever found.

Moreover, a leading expert on the creatures, Dr. Clyde F. E. Roper of the National Museum of Natural History at the Smithsonian Institution in Washington, is mounting a $5 million expedition to observe the giant squid in its habitat. Traveling to the South Pacific near New Zealand, Dr. Roper plans to enter a tiny submersible, dive deep and shadow the beast in the sunless depths, seeking to capture its secrets on film for the first time.

"Our chances are very, very good," Dr. Roper said of the possibility of a deep encounter. "But keep in mind that we had lots of shots at the moon before we got there."

Referring to the cost of the New Zealand foray, he added: "It's a relatively tiny investment when you think of the potential for knowledge and information. We know so little about their biology and behavior."

Dr. Roper has studied the giant squid for decades but, like all other experts, has never seen one alive. Specimens hauled to the surface are usually dead or about to expire, having been battered, squeezed and suffocated in nets full of fish.

The main clues that Dr. Roper and other scientists have followed to locate the animal's habitat are food chains—the progression of who eats whom in nature, from microscopic grazers to mammoth predators the size of apartment buildings. It turns out that the giant squid feeds on certain types of deep fish now being harvested in great numbers and in turn is fed on by sperm whales, giants in their own right that dive down perhaps up to a mile to feast on the boneless leviathans.

Scientists, like hunters following a pack of bloodhounds, plan on tracking the fauna at both ends of this food chain in hopes of discovering the giant squid in the middle, lurking in its dark home. Some experts are a bit leery about doing so, given the beast's 10 large tentacles lined with sucker pads and its reputation for ruthlessness.

"I have a lot of respect for these animals," said Dr. Ellen C. Forch, a fisheries biologist in New Zealand, who for more than 15 years has compiled data on the giant squid. As for the expedition, Dr. Forch said she had no plans to go down in the tiny submersible and preferred to monitor the action from a ship. "I have two small children," she explained. "And they need their mother."

Though very poorly known, and often used as a symbol of humanity's ignorance of the deep, the giant squid already holds a number of records. It is believed to be the largest of all the world's creatures that have no backbones, growing up to lengths of 60 or 70 feet, longer than a city bus. Its hubcap-sized eyes are the largest in the animal kingdom. Some of its nerve fibers are so big they were initially mistaken for blood vessels.

Over the centuries the giant squid has clearly been the inspiration for countless tales of ogres, including the kraken, legendary sea monsters off Norway. Erik Pontoppidan, Bishop of Bergen, in 1753 described an immense sea monster "full of arms" that was big enough to crush the largest man-of-war.

Modern impressions of the giant squid began to form with Jules Verne's *20,000 Leagues Under the Sea,* which was published in Paris in 1871. Drawing on reports of real-life encounters, he depicted the animal fairly accurately anatomically but fabricated its habitat, describing it as living in deep caverns in the sides of submarine cliffs. The cave openings were cloaked by tangles of giant weeds.

As Captain Nemo and his submarine passed one of these dim grottos, a passenger saw a "formidable swarming, wriggling movement" in the weeds. Soon, the submarine and its crew were battling a swarm of giant squids and their writhing tentacles.

In the 20th century, it became clear from sightings, captures and strandings that the giant squid was ubiquitous throughout the sea, though very reclusive. Its scientific name is Architeuthis (pronounced ark-e-TOOTH-iss), meaning chief squid in Greek.

Modern scientists have repeatedly tried to catch the beast and observe it in its deep lair, using nets on long lines, submersibles equipped with bright lights and cameras, and lately robots tied to long tethers—always to no avail. Only fishermen have made successful hauls, always by accident. But lacking the ability or interest to preserve the huge specimens, fishermen typically take a picture or two before throwing the carcass overboard, leaving biologists to lament the lost treasure.

All that began to change in the last decade or so off New Zealand. In a pioneering venture, fishermen and scientists there worked together to develop a series of deep commercial fisheries, going after such exotic fish as hoki, ling and the orange roughy. The focus was Chatham Rise, a rocky plateau the size of Texas that lies hidden a half mile or so beneath the waves and drops off steeply on its sides. Chatham Rise and its flanks were discovered to teem with deep-sea life.

Starting in 1984, as the pace of the deep fishing picked up, commercial fishermen began occasionally hauling up giant squids that were apparently feeding on dense schools of fish at depths ranging between 1,000 and 4,000 feet. A system of reporting was initiated so that Dr. Forch and her colleagues in Wellington, working at what is now known as the National Institute of Water and Atmospheric Research, learned of the catches and often received body parts or whole carcasses to study.

In 1991, the work became more analytical as the government agency commissioned the *Tangaroa*, a 230-foot fisheries research vessel specially built and equipped to probe the deep sea with nets, trawls, sonars and other advanced gear.

In the last few years, Dr. Roper of the Smithsonian has joined the New Zealanders in their work, sifting through the evidence from the accumulating sightings in a hunt for habitat clues. A still unresolved question is whether the beast spends most of its time swimming through inky midwaters or near the bottom.

Since December, a run of landings has thrown the field into a high state of excitement. The Australians caught a large female near Tasmania. And on Chatham Rise and a more westerly site, the New Zealanders caught two females, 26 feet and 13 feet long, as well as a prize male, the physically smaller of the two genders and for some reason extremely rare. It was 20 feet long and discovered at a depth of 1,000 feet.

"That's very shallow," Steve O'Shea, a marine scientist at the National Institute of Water and Atmospheric Research in Wellington, said in an interview. "We expect them to be around 1,000 meters," or 3,280 feet below the surface.

Some ecologists fear that deep fishing is unsettling the squid's diet and domain, possibly forcing the animal into shallower waters. But scientists say they cannot resolve the issue because information on deep ecology is so scarce.

Mr. O'Shea, who is in charge of collecting giant squid data throughout New Zealand waters, from both government and commercial vessels, said the new specimens were in beautiful shape. In the past, dead animals have often been a shambles, hacked into pieces by fishermen or so bruised and abraded that parts were unrecognizable.

"They're intact!" Mr. O'Shea said of the new specimens. "No tentacles are torn off."

As for the overall hunt, he added, "We're getting a very tantalizing glimpse of where we expect to find these things."

The scientific team consisting of the New Zealanders, Dr. Roper and his Smithsonian colleagues believe that the time is ripe to make a concerted push to observe the giant squid in its habitat.

"Seeing a giant squid would be the ultimate," said Dr. Forch, the fisheries biologist. "Naturalists for a long time have been going after all sorts of exotic things that are easy to get to. But this is very remote and elusive. It's just out of reach."

If enough money can be raised, the team plans to conduct the hunt between late November and February, which is summertime in New Zealand and a season when the giants are frequently found. In a two-pronged attack, the team first plans to use the research vessel *Tangaroa* to make preliminary searches for deep fish types and densities. Then, its specialists will deploy binoculars up top and underwater microphones below to track and listen to sperm whales as they dive into the depths to eat squid, hopefully guiding the scientists to the lair.

If all goes as planned, team members plan to send a robot down to inspect the area and then to dive personally into the inky darkness in the *Johnson-Sea-Link,* a submersible operated by the Harbor Branch Oceanographic Institution in Fort Pierce, Florida.

Outfitted with robotic arms, lights and video cameras, the submersible is made of a single large acrylic sphere for maximum visibility and can carry up to four people to depths of 3,000 feet, well within the bounds of the beast's apparent home. Teaming up with the scientists will be a crew from *National Geographic,* which plans to televise the encounter.

"If we find one and film it, that would be absolutely spectacular," said Dr. Roper, who seems to have no fear that the submersible will be wrapped in giant tentacles and crushed or crippled.

"A few minutes of film would show a lot," he said. "Seeing a giant squid from a submersible would open a new world of understanding."

—WILLIAM J. BROAD, February 1996

Violent World of Corals Is Facing New Dangers

DENSELY ABLOOM WITH A WEALTH OF LIFE unsurpassed for bizarre beauty, coral reefs seem to exist in a state of dreamy tranquillity. Not so.

The reef is in truth a realm of violent struggle and constant disruption. Coral colonies wage unrelenting chemical warfare on each other, their polyps stinging, dissolving and poisoning each other. Bigger reef creatures savage large chunks of colonies and fill the water with toxins. Sooner or later, an irresistible force like a hurricane or a change in sea level lays waste the whole teeming ecosystem and the corals must rebuild.

But rebuild they do, and this resilience is at the heart of a dispute among marine biologists over the contribution of human activity to the stresses on coral ecosystems.

One school of thought holds that corals worldwide are now in serious peril because of human assaults like global warming, overfishing, pollution and physical destruction of reefs by fishermen and tourists.

An opposing school holds that while some reefs are indeed in big trouble, many others remain pristine and even the damaged ones have adapted in the past to natural forces at least as destructive as human activity.

Any serious threat to corals would be an ecological tragedy. The biological diversity of coral reefs compares with that of tropical forests. Reefs themselves, built from the calcified skeletons of polyps, are the largest structures created by life. The many strange toxins evolved by reef denizens for their biological warfare hold considerable promise as treatments for various human diseases.

For all these reasons, there has been a heightening of scientific interest in corals of late, including an outpouring of research on coral ecology and chemistry.

Tentacles

Individual living coral

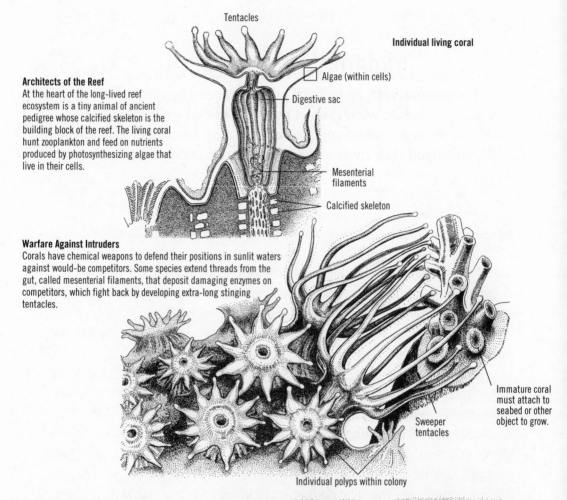

Algae (within cells)

Digestive sac

Architects of the Reef
At the heart of the long-lived reef ecosystem is a tiny animal of ancient pedigree whose calcified skeleton is the building block of the reef. The living coral hunt zooplankton and feed on nutrients produced by photosynthesizing algae that live in their cells.

Mesenterial filaments

Calcified skeleton

Warfare Against Intruders
Corals have chemical weapons to defend their positions in sunlit waters against would-be competitors. Some species extend threads from the gut, called mesenterial filaments, that deposit damaging enzymes on competitors, which fight back by developing extra-long stinging tentacles.

Immature coral must attach to seabed or other object to grow.

Sweeper tentacles

Individual polyps within colony

Is Shape Destiny?
The vertical growth of a branching coral can be an asset in competition for space, for once it gains a foothold it can deprive competitors of sunlight. But its shape makes it vulnerable to destruction in storms. Huge coral colonies with mound-like shapes can lose access to sunlight, yet their resilience in storms lets them live for centuries.

Elkhorn

Brain coral

Dimitry Schidlovsky

Coral colonies, each composed of innumerable tiny, tentacled polyps, take the forms of trees, shrubs, fans, plates and huge boulders. The phantasmagoria of shapes creates a habitat for other marine creatures like fish, lobsters, sponges, mollusks, octopuses and sea anemones.

Competition with no quarter given is the rule in this interdependent but mutually hostile world. When polyps in one colony come face-to-face with another in a constant competition for scarce space, they expand their bodies to engulf their rivals and exude digestive juices that turn the competitors to jelly.

As a countermeasure, polyps in the second colony grow "sweeper tentacles" studded with special stinging organelles that "zap the neighbors," says Dr. Judith Lang, a reef ecologist at the Texas Memorial Museum at the University of Texas.

Still other polyps enshroud their enemies in a sticky mucus that dissolves the tissues. Combat with toxins is also thought to be a common mode of warfare, though the toxins have proved hard to pin down in actual use.

Coral reefs may be one of the most naturally poisonous environments on earth. One study found that 73 percent of 429 species of exposed invertebrates commonly occurring on Australia's Great Barrier Reef were toxic to fish.

Although much remains unknown about how the toxins are used, they have "tremendous potential" as pharmaceuticals, said Dr. Drew Harvell, a coral reef ecologist at Cornell University. For example, Dr. Harvell said, some toxins produced by soft corals "are anti-inflammatory agents, some have potential as being effective against AIDS and some as anti-cancer drugs, and that's just one class of organisms."

Because of this great potential, the reefs have recently become prime prospecting grounds. Many potentially useful compounds have been discovered and are now undergoing further testing, said Dr. David J. Newman, a chemist in the National Cancer Institute's Natural Products Branch. The center has been collecting about 1,000 samples of coral-reef organisms a year for the last five years. The prospecting effort is still young; the journey to market for any drug derived from natural sources typically takes 5 to 15 years.

In nature, the coral toxins may play an indirect but key role in the reef ecosystems' resilience in the face of disturbance. Dr. Robert Endean and Dr.

Ann Cameron of the University of Queensland in Australia, who have long studied the Great Barrier Reef, postulate that extensive boulder-like coral colonies have been able to exist continuously for hundreds and even thousands of years because they are so successful in using toxins to ward off predators.

Because the shape and mass of these long-lived "persister" colonies protect them from the storms that devastate more fragile corals, Dr. Endean theorizes, they form the staunch backbone of the reef ecosystem. They must be protected from destruction by humans at all costs, he warns.

Scientists generally agree that humans are putting serious stress on at least some coral reefs. Sediment from dredging and agricultural runoff, sewage, chemical pollution, dynamiting of reefs by fishermen, damage from boat anchors and scuba divers, shell-collecting, mining of reefs for calcium deposits and overfishing that shatters ecological checks and balances all play a role.

"I don't know how to state it in strong enough terms," Dr. Harvell said of the destruction. There has been especially widespread degradation in the Philippines and other Southeast Asian waters. "It's not just that there aren't any fish on these reefs," Dr. Harvell said. "That's bad enough, but there aren't any corals on many. You don't want to panic everybody that the world is falling apart, but some places are."

But there is disagreement about the extent of the degradation, about the ability of the reefs to withstand human-induced stress and about whether some of the damage is caused by humans at all.

Some scientists contend that the threat is global. But "very few people have looked at reefs all over the world and have the big picture," said Dr. Richard Grigg of the University of Hawaii, who edits *The Journal of Coral Reefs*. While it is true that the human impact has been heavy in Southeast Asian and Caribbean waters, he said, reefs in wide areas of the Pacific remain pristine. "There is no one answer," he said. "You have to look at it case by case."

Some threats seem more general, but there is doubt as to whether humans are responsible. In recent years, corals in several parts of the world have turned white and died. Coral polyps are largely nourished by carbon compounds produced by algae in their cells. The algae also give the polyps their brilliant colors.

Some scientists testified before Congress that because the algae die when the water gets too warm, human-induced global warming was a likely cause of the bleaching. But many other scientists say that while warmer water does cause coral bleaching, it is too soon to tell whether the oceans are warming as a result of waste industrial gases that trap heat in the atmosphere. The warmer water, they say, could be a natural phenomenon.

Some scientists also say human activity is to blame for widespread damage to corals by a population explosion of the highly predatory crown-of-thorns starfish. The starfish's own natural predators were decimated by overfishing, it is contended. Other scientists, noting that the starfish population is just as out of control on pristine Pacific reefs, argue that natural fluctuations in animal populations are at work.

But since some human-induced damage is indisputable, does it portend disaster?

One body of thought, with which Dr. Endean has been associated, sees reef ecosystems as being so delicately adjusted to natural conditions that disturbance by humans could impoverish them indefinitely. An opposing body says that far from being in fine balance, reefs are always recovering from some natural disturbance or other, and that there is little effective difference between natural disruptions and those caused by humans. "Nature can be and often is more perverse than man," Dr. Grigg said.

For example, in 1980 Hurricane Allen smashed vast stretches of tree-like staghorn corals in the Caribbean, causing a total collapse of this formerly dominant ecosystem element along the north coast of Jamaica. And at the end of the last ice age, most of the world's corals died when rapidly rising sea waters shut out the sunlight on which their life-giving algae depend. Most of today's reef ecosystems developed after that. The oldest are about 8,000 years old.

The question of coral reefs' hardiness depends on what time scale is considered, said Dr. Jeremy B. C. Jackson, a senior scientist and coral reef ecologist at the Smithsonian Tropical Research Institute in Panama. Corals are among the most ancient of animals. On the longest scale, they have also been among the most fragile; living reefs have disappeared from the planet entirely as part of more general extinctions of life in the remote past. But for the last million years, despite temporary setbacks like the one at the end of the last ice age, they have generally been remarkably stable.

On the scale of a human lifetime, Dr. Jackson said, it may look as if "the sky is falling, and in some cases with good reason." That is, he said, if a hotel owner in Grand Cayman or Jamaica suddenly finds that the coral reefs that attract tourists have become trash, things look pretty bad and unstable.

From this viewpoint and that of other people who have much to lose from reef destruction, say scientists on all sides of the question, what humans are doing to the reefs is serious. The longer the stress continues, the longer natural recovery is delayed.

But Dr. Grigg, no adherent of doomsday thinking, posed this question: "How many corals are extinct from either natural or human-induced changes in the last couple of hundred years?" And then he answered: zero.

Even so, he said, a new coral-reef doomsday is not unthinkable. Although corals over the last several million years have survived "all manner of catastrophes," including meteor and comet impacts and some 20 ice ages, he recently wrote, this proof of robustness "may hold little consolation" in the face of a vast and rapid expansion of human population. Unless the expansion is reined in, both corals and people themselves could be threatened; and that, he wrote, may be the real truth spoken by the reefs.

—WILLIAM K. STEVENS, February 1993

Freshwater Mussels Facing Mass Extinction

THE MEMBERS OF THE MOST WIDELY ENDANGERED family of organisms in the United States have neither fur nor feathers, and hardly any friends. Politically as much as anatomically, they are truly faceless. They are freshwater mussels, and they are in the midst of what experts call a mass extinction.

More species of the family of freshwater mollusks known as Unionidae are among the candidates being considered for protection under the Endangered Species Act than any other taxonomic family. Many are already protected, while others have slipped into extinction.

Of nearly 300 species native to the United States, half are in serious trouble: About 20 are considered extinct, about 60 are listed as threatened or endangered and about 70 have been proposed for listing. When *Fisheries,* the journal of the American Fisheries Society, published a survey of the entire fauna in 1993, it listed only 70 species as stable.

"Everybody talks about biodiversity in the tropical rain forests," said Dr. Arthur E. Bogan, a research associate in invertebrate zoology at the Carnegie Museum of Natural History in Pittsburgh. "But you don't have to go to the tropics to see a major extinction event." In a 1993 article in the peer-reviewed *Journal of the American Society of Zoologists,* he said that the entire family of mussels was "poised on the brink of a major and widespread extinction."

The extinction of the mussels has been documented for decades, and is not in dispute among malacologists. They understand, too, why it is happening. Habitat is fragmented as streams are dammed. It is depleted by dredging and gravel mining. It is destroyed by siltation from streamside logging and agriculture, and it is poisoned by discharges of sewage, toxic wastes and polluted runoff.

Typical of the human activities that endanger the mussels is dam building. Few rivers are left to flow completely naturally; many are dammed up and down their lengths. And because of the peculiarities of mussel reproduction, damming can cause lethal harm to a species that may live only in a single stream.

A young mussel passes through a larval stage where it matures while attached to the gills or fins of a host fish. Mussels have evolved a variety of mechanisms to tempt fish into serving as obligate hosts; some display attractive, bait-like body parts to draw fish nearby, and others package the larvae themselves to look like worms or minnows. Dam a creek and the host fish may be unable to swim up- or downstream to reach the mussels and help them reproduce and disperse. Cut off from the obligate hosts, a local population of mussels becomes functionally extinct, even though individual mussels may survive into old age, which in some species is several decades.

The American diversity of mussels is extraordinary; they are the beetles of Southeastern streams. Their names are a folk poet's playground: Waccamaw fatmuckets and fuzzy pigtoes, pocketbooks and snuffboxes, purple bankclimbers and Tennessee heelsplitters.

"The United States is home to approximately one third of the world's freshwater mussel species," said a report by Dr. Lynne Corn, a natural resources policy specialist at the Library of Congress. "Freshwater mussels are regarded as important indicators of aquatic ecosystem health because they are very sensitive to changes in water quality. As such, these seemingly inconsequential organisms are assigned tremendous ecological importance by many freshwater biologists."

Mussels once provided shells for buttons; now they are mainly the source of blanks inserted as seed for pearls grown in oysters. Among their ecological functions are filtering water and, of course, serving as food for other species. A purple bankclimber can reach nine inches in length. "That's a sit-down meal for a possum," said Dr. Richard Neves, a malacologist at Virginia Polytechnic Institute.

It is in the Southeast that mussels are most diverse—and most endangered. Their waning is a sign of serious problems in the whole freshwater aquatic ecosystem of the region, scientists say.

Two scientists reported in the August issue of the journal *Conservation Biology* that the rivers and streams of Alabama contain 60 percent of known mussel species—and 69 percent of the state's species are either extinct, endangered, threatened or of special concern, in diminishing order the categories of peril used by wildlife experts. Similar trends are found in the state's freshwater fish, gill-breathing snails and freshwater turtles, said the article, by Dr. Charles Lydeard and Dr. Richard L. Mayden of the University of Alabama in Tuscaloosa.

Calling the state's waters "an extraordinarily diverse and endangered ecosystem," they wrote, "Extinction is known to have occurred, species are known to be endangered and threatened, and the ecosystems contained within this hot spot face daily abuse."

If relatively little has been done over the years to allay the problems that face the mussels, it is an illustration of the problems of controlling harmful human activities across whole ecosystems.

In the review of the fauna published in *Fisheries* in 1993, Dr. James Williams, a researcher with the National Biological Service in Gainesville, Florida, called on agencies to try to protect whole ecosystems rather than focusing on one species at a time, but "grimly" admitted that many species "obviously need emergency attention or they will perish."

But there is little of an emergency nature in recovery plans for the mussels or for their ecosystems, as is shown in a draft recovery plan for several imperiled aquatic species in the Mobile River basin, which drains much of Alabama and nearby states. The plan is currently being circulated for public comment by the Fish and Wildlife Service. Among the species covered by the plan are three small fish, a snail and 11 mussels.

Like other studies, this one finds that the diversity of fish and mollusks in the drainage is "unmatched in North America," but that "the basin is experiencing a high rate of biotic extinction—a rate unparalleled in the continental United States."

But the plan acknowledges that the human footprint on the landscape is irreversible, and that "humans and their activities are also components of the ecosystem."

"An informed public adopting voluntary changes in land and water stewardship is the only practical way to support both humanity and the basin's unique aquatic community," it says. And recovery of the mussels to the point of being removed from the endangered species list "is unlikely in the near future because of the extent of their decline, their apparent sensitivity to common pollutants and continued impacts upon their habitat."

—JOHN H. CUSHMAN, JR., October 1995

Earliest Samples of
Royal Purple Found

PURPLE HAS LONG BEEN THE COLOR of royalty and high ecclesiastics, and through time the most regal of the purples came from a dye made from a gland of certain Mediterranean mollusks. The poet Robert Browning spoke of it as "the dye of dyes." In the Roman Empire, the dye's rich hue and colorfast properties were so valued, and its production so time-consuming, that the extract was literally worth its weight in gold.

The Phoenicians became proficient in the production of royal purple and in the first millennium B.C. engaged in a brisk trade in dyed fabrics. Wherever the seafaring Phoenicians went they set up dye factories, leaving mounds of mollusk shells to fascinate and inform future archeologists. Their name may even derive from a root meaning "purple."

Scientists have now come up with chemical evidence that people in the late Bronze Age, several hundred years before the Phoenicians, were also busy making the dye out of such mollusks as *Murex brandaris* and *Murex trunculus*. These marine snails are often called whelks.

In an analysis of a purple-colored accumulation in an earthen jar found in Lebanon, chemists at the University of Pennsylvania and E. I. Du Pont de Nemours & Company identified what they said were the earliest known samples of the famous dye. Pottery at the site, at Sarepta, midway between Tyre and Sidon on the Mediterranean coast, was dated between 1300 and 1200 B.C. The Phoenician culture rose in the beginning of the Iron Age, around 1000 B.C., replacing the Bronze Age Canaanites in that region.

Other evidence, from shell middens and references in documents, suggests that the dye industry was thriving as early as 1500 B.C., possibly in Crete and Syria.

The jar that was the object of the analysis was among many artifacts excavated in the 1970's by James B. Pritchard of the University Museum of Archeology and Anthropology at the University of Pennsylvania in Philadelphia. The jar was more than six inches high and had a spout. A nearby heap of mollusk shells led Dr. Pritchard to surmise that this had been a vat for processing purple dye.

Confirmation was provided by Patrick E. McGovern, an archeochemist at the University Museum, and Rudolph H. Michel, a retired Du Pont chemist. Their analysis, conducted at Du Pont in Wilmington, Delaware, the University of Delaware and the University Museum, was described in a recent issue of a journal published by the museum.

The two chemists submitted the pot's dark purple deposit to X-ray spectroscopic and infrared-radiation examination. The X-ray analysis revealed unusually high levels of bromine, a strong clue. The infrared test established more specifically the presence of dibromo-indigo, the known ingredient of the ancient purple dye. Further tests ruled out the possibility that this was a modern synthetic dye.

In an additional test, Dr. McGovern and Dr. Michel took scrapings from the deposit and produced a solution that exhibited the chemical behavior of the dyes made from two Murex species, either trunculus or brandaris. Most of the shells at the site were trunculus.

Dr. McGovern and Dr. Michel concluded: "The combined evidence from the spectroscopic and chemical investigations leaves no doubt that the major component of the purple deposit on the Late Bronze Sarepta sherd is 6,6'-dibromoindigotin. This is the earliest chemical confirmation of the ancient dye."

From earlier studies scientists have determined that the source of the dye is the hypobranchial gland in the Murex. They believe the dye might be ejected by the mollusk as a defensive mechanism, much like the black ink of the octopus or squid.

The best description of the dye-making process was given by Pliny the Elder in the first century A.D. In his *Natural History,* Pliny said the mollusks were caught in baited wicker baskets and their flesh, containing the gland, was removed and thrown into open holes where they decomposed for three days. The strong odor was compared to rotting garlic.

The mass was then placed in stone or lead vats and, with water added, gently heated for 10 days. It took 8,000 pounds of mollusk flesh to make about 500 pounds of the dye-producing substance.

The dyes produced various colors, from blue to violet to black-purple, depending on the mollusk species and quantities used. Pliny noted that the most desirable color was that of congealed blood.

—JOHN NOBLE WILFORD, March 1985